GREENWICH TIME

AND THE LONGITUDE

GREENWICH TIME

AND THE LONGITUDE

Derek Howse

PHILIP WILSON

NATIONAL MARITIME MUSEUM

in association with

ATKEARNEY

© 1997 Derek Howse,
First published 1980 as *Greenwich Time and the Discovery of the Longitude*, by
Oxford University Press. This edition published by
Philip Wilson Publishers Limited
143–149 Great Portland Street
London W1N 5FB

Distributed in the USA and Canada by
Antique Collectors' Club Limited
Market Street Industrial Park
Wappingers' Falls
NY 12590
USA

Library of Congress no.: 97-061678
ISBN 0-85667-468-0 (Hardback)
ISBN 0-948065-26-5 (Paperback)

Managing Editor: Lynn Bryan
Editors: John Gilbert and Tim Ayers

Designed by Chapman Bounford & Associates
Printed and bound in Italy by Artegrafica S. P. A. - Verona

Front Cover. The modern Prime Meridian of Longitude at the Old Royal Observatory, Greenwich, with (behind) a detail from General Roy's survey of 1784, showing the Bradley Prime Meridian of 1750. This is in fact 5.79 m (19 ft) to the west (right) and is still the basis of Britain's land maps.

Back Cover. The Old Royal Observatory at night and its Shepherd Gate Clock, 1852, the first to show Greenwich Time to the public.

Frontispiece. The domes of Greenwich: left to right, the time-ball on Flamsteed House, the Altazimuth dome (with Halley's Comet wind vane), the 28-in Great Equatorial Telescope dome. Inset. The Old Royal Observatory, Greenwich, from the north.

Page 11. The Bradley Transit Room at Greenwich. Between 1750 and 1850 the Greenwich meridian was defined by the transit instrument in this room, above the chair.

Photographic Credits
The Author and Publisher thank the following for photographs and permission to reproduce them: in London, the Bridgeman Art Library (p. 92), Trustees of the British Library (pp. 16, 22, 30, 122) and of the British Museum (p. 60), the National Portrait Gallery (p. 33), the Royal Naval College, Greenwich (p. 37), the Royal Society (p. 69), the Science Museum/Science and Society Picture Library (pp. 65, 72, 77, 93) and the *South-east London and Kentish Mercury* (p. 113); in Paris, Conservatoire National des Arts et Métiers (p. 79), Observatoire de Paris (pp. 132, 189); and the Bodleian Library, Oxford (p. 23), Canadian Pacific Corporate Archives (p. 131), Dover Museum (p. 102), Hewlett Packard (p. 174), Marconi Ltd (p. 159), New Zealand Historic Places Trust (p. 84), B. and E. Nicholls (p. 54), Powerhouse Museum, Sydney (p. 84), the Royal Greenwich Observatory (pp. 86, 91, 101, 165, 166, 168, 170, 173, 188) and Humphry M. Smith (p. 177). All other photographs were supplied by the National Maritime Museum.

Contents

Celestial chart of the southern hemisphere from Renier Ottens' Atlas, 1730. The chart is copied from that of John Flamsteed, first Astronomer Royal at the Greenwich Observatory, which is shown in the top left-hand corner. The chart shows three other European observatories: Hesse-Cassel, Berlin and Hven (also shown).

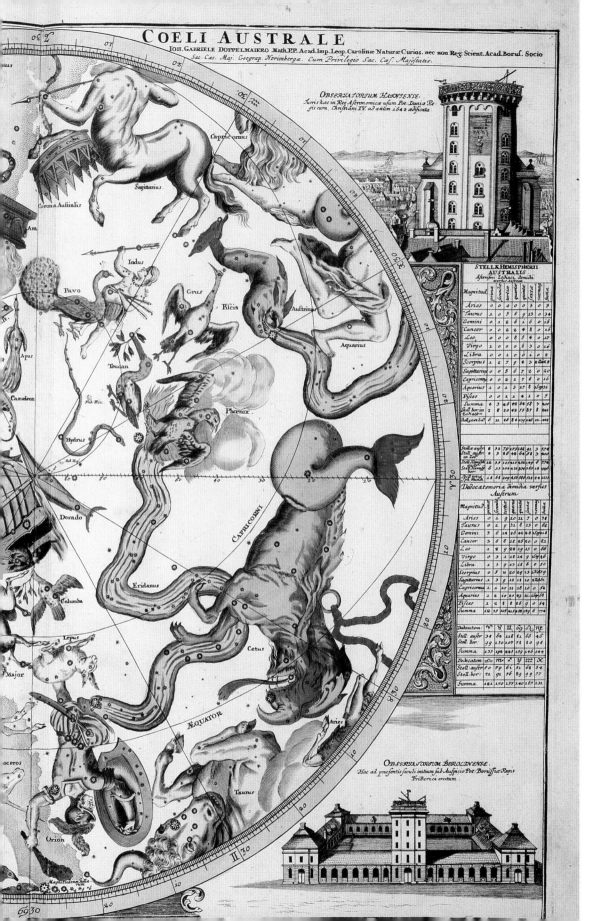

Acknowledgements

This is a new edition of a book first published in 1980 by the Oxford University Press under the title *Greenwich Time and the Discovery of the Longitude*. Among those whom I had the pleasure of thanking for their help in that first edition, I am most happy to be able to thank once again, for their help in this new edition: Mr Humphry M. Smith (one-time 'Mr GMT'), Professor Sir Francis Graham-Smith, Mr Andrew Murray and Mr Leslie Morrison, all then of the Royal Greenwich Observatory at Herstmonceux; and Mlle Suzanne Débarbat of Paris Observatory.

Among the many new friends and other contributors who have helped in this edition, I must make special mention of Philip Banks and Jackie English at A. T. Kearney; John Chambers of the National Physical Laboratory; Elaine Marland at the New Zealand Historic Trust; Carl Calvert of Ordnance Survey; Lynn Bryan, Anne Jackson and Louise Tucker at Philip Wilson Publishers; Dr Jasper Wall and Dr Andrew Sinclair of the Royal Greenwich Observatory, now at Cambridge; Adam Perkins of the RGO Archives, now in the Cambridge University Library; Dr Paul Murdin, of the Particle Physics and Astronomy Research Council; Group Captain David Broughton of the Royal Institute of Navigation; Keith A. Pickering of Maryland, USA; Dr J. I. Hunt and Howard Block of London; Mike Feist; Jonathan Betts, Maria Blyzinsky, Tina Chambers, Jane Costantini, Dr Gloria Clifton, Peter Robinson and Catherine Sones of the National Maritime Museum.

Most of the illustrations in the 1980 edition are repeated here, with others, many now in colour, though redesign has necessitated changes in their format and length of captions.

Despite these changes, all significant information relating to the illustrations is still present in the captions and the main text. The only large omission is of Appendices III to VI of 1980 (on mechanical and precision clocks, an 1861 list of time-balls and a delegate list of the 1884 Meridian Conference). These have been replaced by a new Appendix III by Carl Calvert, for which I am much obliged.

I am grateful to the National Maritime Museum, Greenwich, and its Head of Publications, Dr Pieter van der Merwe, for the initiative of organizing this edition, and join the Museum in warm thanks to A. T. Kearney Limited, whose support has made it possible.

Finally, I am most grateful to the present Astronomer Royal, Sir Martin Rees, for doing me the honour of writing the Foreword.

Derek Howse
1997

Preface

As an international company that advises others how they should best deploy time and other resources, we are delighted to be associated with this Millennium edition of the 'official history of Greenwich Time'. We find it fascinating how a story which began in the 1670s as a scientific project to improve navigation became, in the 1850s, one for co-ordinating the general timekeeping needs of Britain, and eventually of the whole world. As so often happens, research in one direction delivers innovative long-term benefits in others.

The later twentieth century has seen the world shrink to become an electronic 'global village', in which neither time nor distance are barriers to the tide of information which now dominates our lives. That tide, like the sea itself, still largely ebbs and flows to the rhythm of the turning globe, and to the beat which the Prime Meridian of Longitude at Greenwich has set on our world time system since 1884.

The story told in this book is a heartening one: overall its triumphs have been achieved through knowledge, reason and co-operation, rather than conflict. If Greenwich Time survives to mark the dawn of the year 3000, we hope that its unifying message will continue to show the intervening centuries an example of what peaceful scientific endeavour can achieve for the common good.

Jan-Willem Broekhuysen
Managing Director
A. T. Kearney Limited

Foreword

Astronomy is the oldest quantitative science. The need for exact calendars, safer navigation and accurate timekeeping has stimulated the subject from ancient times. Government support for English astronomy dates from 1675, when Charles II's initiative led to the foundation of the Royal Observatory at Greenwich. The 'zero-longitude' meridian that slices through it has unique status, and permanently commemorates its preeminent historical role.

John Flamsteed, the Observatory's first director, embarked on a programme to map the stars – a prerequisite for measuring longitude by the so-called 'lunar method'. In 1714, Parliament set up the Board of Longitude – the first scientific 'quango' – with a budget to foster new ideas, and to provide a very handsome prize for whoever devised an accurate technique for measuring longitude. The lure of this bounty stimulated improvements in clockmaking that led to Harrison's celebrated chronometers – this saga is chronicled here, as are the (often bizarre) unsuccessful approaches to the longitude problem.

In the nineteenth century, the railways and telegraph heightened the need for precise and coordinated timekeeping; by about 1870, Greenwich Time became the standard throughout the United Kingdom. The international discussions on 'time zones', and on worldwide standards, were more contentious. After prolonged negotiation, the 1884 Meridian Conference, held in Washington, settled on the Greenwich Meridian as the zero of longitude. Right up to the present day, the need for ever greater accuracy in measuring Greenwich Mean Time (later to be internationalized as Universal Time) has continued to stimulate innovation at the forefront of technology.

Derek Howse has unique credentials to tell this story. He combines the authority of a distinguished scholar with the lively style of a practised writer. As the Millennium approaches, interest will focus on the Prime Meridian as never before. This fine new edition of *Greenwich Time* brings the story up to date.

Sir Martin Rees, *Astronomer Royal*

Introduction

I. That it is the opinion of this Congress that it is desirable to adopt a single prime meridian for all nations, in place of the multiplicity of initial meridians which now exist.

II. That the Conference proposes . . . the adoption of the meridian passing through the centre of the transit instrument at the Observatory of Greenwich as the initial meridian for longitude . . .

V. That this universal day is to be a mean solar day; is to begin for all the world at the moment of mean midnight of the initial meridian, coinciding with the beginning of the civil day and date of that meridian . . .[1]

Today, the name Greenwich is known and used by men and women of every race and creed. This is the result of the resolutions quoted above, resolutions taken on 22 October 1884 at the end of the International Meridian Conference which met in Washington, DC, 'for the purpose of discussing, and, if possible, fixing upon a meridian proper to be employed as a common zero of longitude and standard of time-reckoning throughout the whole world'.[2]

This book tells the story of Greenwich time. It tells how, three hundred years ago, it was used only by those who lived and worked in Greenwich, perhaps a few hundred people, the most important from our point of view being two astronomers in the newly founded Royal Observatory in Greenwich Park; how, two hundred years ago, it began to be employed by seamen of all nations who used the newly published British *Nautical Almanac*, an annual publication whose data were based on the Greenwich meridian; how, a hundred years ago, improved communications made it desirable to co-ordinate the time kept worldwide, resulting in the system of standard times used today, based on the Greenwich Meridian; how Greenwich Mean Time has today become Universal Time, the basis of the time employed for both domestic and scientific purposes all over the world – and why it was Greenwich time (and not, say, Paris time) that was chosen as the world's Universal Time.

The need for such Universal Time, a time-scale of worldwide application and available to the world at large, has, however, arisen only quite recently. Since the earliest times Man has regulated his daily activities by the Sun – by its rising, its culmination (noon or midday), and its setting. Indeed, in many early civilizations, the day was divided into twelve 'hours of the day' (sunrise to sunset) and twelve 'hours of the night' (sunset to sunrise), which were of different length. As summer days are longer than winter days, not only were the day-time hours of different length from the night-time hours (except at the equinoxes in March and September), but they also varied according to the seasons. Travellers discovered there were also variations according to how far one was from the equator.

This system of 'unequal hours', based on local sunrise and sunset, with all its apparent complications, worked well enough for most domestic and business purposes except in the more northerly places where the discrepancy between night and day might be considerable. Indeed, it was still in use by some people in Italy until the fifteenth century. Greek astronomers, however, had divided their day into twenty-four hours of equal length and their successors of all nations continued this tradition.

Whatever system of hours was used, however, the time kept by the ordinary person was *local* time, as shown by a sundial at the place where he was. To that ordinary person it was of no consequence that a place to the east or west of him actually kept different time by the Sun; that (because of the rotation of the Earth) when it was noon in London, it was 11.44 in Plymouth but 12.05 in Norwich; or, put another way, that noon (when the Sun was due south) occurred in London 16 minutes before Plymouth but 5 minutes after Norwich. When Man's fastest speed of travel depended upon the horse, what did these differences matter? In any case, only in the last few hundred years has the keeping of time to any sort of accuracy been of any consequence in daily life, and then generally only in towns.

Although the ordinary person might have no need for accurate time, there were some – geographers, map-makers, astronomers, travellers and navigators – who did have such a need, not to find time for its own sake, but because, through time, they might be able to find, for example, the difference of longitude between places. How to do it was simple – in theory. Say you wanted to find the difference of longitude between London and Plymouth: if you measured the exact local time in London – say noon – and had some way of discovering the exact local time *at the very same moment* in Plymouth – say 11.44 – then the difference – 16 minutes – gave you the difference in longitude between the meridians of London and

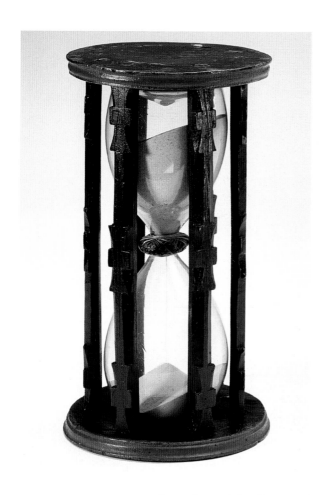

Hourglass, c. 1740. These were used for general shipboard timekeeping, until pocket watches and chronometers became common in the 1800s.

Plymouth, 16 minutes of time being equivalent to four degrees of arc. Because Plymouth noon occurs after London noon, Plymouth must be west of London, by 4° (the problems of finding longitude are dealt with more fully in Appendix I).

But the gap between theory and practice was a large one. How could you find simultaneously the local times of an event as recorded at two widely separated places? Of course, there might be other methods of finding differences of longitudes – actual measurement of distances on the ground, or measuring the Earth's magnetism, for example – but the astronomical method, by finding the difference of times, seemed always to offer the best chance of success. The basic concept was known to the Greek astronomer Hipparchos by *c.* 180 BC but the accurate determination of longitude had to wait until the 1650s on land, and the 1770s at sea.

The early part of this story of Greenwich time was thus entirely bound up with efforts to find a practical method of finding longitude, particularly at sea where it was made all the more difficult by the motion of the ship and the length of the voyages. This was no abstract concept: it was a highly practical necessity, as will be made clear.

But the revolution in navigational methods today brought about by developments in electronics – from Radar, through Loran and Decca, to today's satellite Global Positioning System (GPS) – all of these depend upon the very precise measurement of time and time intervals, not just to the nearest second as in astronomical navigation, but to the nano-second, one thousand millionth of a second, the time it takes light to travel 1 foot. There are as many nanoseconds in 1 second as there are seconds in 32 years.

Although Universal Time is now coordinated from Paris, the world's prime meridian for longitude and time still passes through the old observatory buildings at Greenwich, and the world's time signals (Coordinated Universal Time, or UTC) are not allowed to depart more than 1 second from the old GMT.

'The centre of Time and Space': the Prime Meridian of the world, as marked in front of the Airy Transit Circle which defines it, in the courtyard of the Old Royal Observatory, Greenwich. The exact point from which the world's time is measured is at the cross-hairs of the Transit Circle – Latitude 51° 28' 38.2" North, Longitude 000° 00' 00".

The old buildings of the Royal Observatory in Greenwich Park now form part of the National Maritime Museum.

ONE

Seeking the longitude

300 BC – AD 1675

Opposite.
A detail of the cross-staff in use (see p. 22).
In the first published description of the lunar-distance method of finding longitude, in 1514, Johann Werner of Nuremberg said that the instrument to be used for measuring the angular distance in the sky between the Moon and a star should be the cross-staff, seen here for measuring a lunar distance. From the title page of Peter Apian's *Introductio Geographica Petri Apiani in Doctissimus Verneri Annotationes ...* (Ingolstadt, 1533).

THE concept of geographical latitude and longitude for defining positions on the Earth's surface had probably come into use in ancient Greece before 300 BC, but not in the way we think of today – of latitudes so many degrees north or south of the equator and longitudes so many degrees east or west of some chosen meridian. In Hellenistic times these quantities were usually – but not always – thought of in terms of time, in the number of daylight hours on the longest day of the year in a particular latitude, in the difference in local time between two places for longitude differences east or west. So far as we know, the first person to offer a mathematically clear theory of geographical latitude and longitude was Claudius Ptolemy (c. AD 100-165) who in his *Geography* created a consistent grid of coordinates, reckoned in degrees instead of the traditional time co-ordinates, with latitudes measured from the equator and longitudes from the westernmost point of the known world, the 'Fortunate Isles'.[1]

Ptolemy, who spent most of his working life in the great library of Alexandria, produced two major works, the impact of which continued to be felt until the seventeenth century. The first of these was his 'Great Collection' or *Megale Syntaxis*, better known as the *Almagest*, a synthesis of all that was best in Hellenistic mathematical astronomy, which continued the work of Hipparchos and Apollonios and added much material of his own. The second was his *Geography*, a gazetteer and atlas of the known world in eight books, giving geographical positions for many thousands of places. In a long introductory treatise on map-making Ptolemy dealt with the finding of geographical positions, discussing in particular the methods advocated three hundred years before by Hipparchos of Nicaea in Bithynia (c. 190-120 BC), whose proposals for finding longitude are particularly relevant.

In Hipparchos's time it was realized that finding differences of longitude would be possible if the same event could be observed (and the local time of that event measured) in each of the two places concerned. Hipparchos suggested making use of the eclipses of the Moon for that purpose because the entrance of the Moon into the Earth's shadow is something which occurs at precisely the same moment for all observers regardless of their position on the Earth.

What Hipparchos failed to explain was how the local time of each place should be found. Because the Sun *must* be below the horizon during a lunar eclipse, a sundial cannot be used directly. Various possibilities are discussed in Appendix 1. But there were other difficulties with the lunar-eclipse method. Eclipses are comparatively rare in any one place, sometimes two or three a year, sometimes none at all – so finding the longitudes of a large number of places takes a long time. In order that no opportunity should be missed, Hipparchos is said to have compiled a list of future eclipses for the succeeding six hundred years.

Another difficulty with the lunar-eclipse method is that defining a particular point in an eclipse is not easy. The 'beginning', the 'middle' and the 'end' of an eclipse can mean different things to different people, so that errors of several minutes of time – and therefore of degrees or more of longitude – can creep in from that cause alone. Nevertheless, Hipparchos's eclipse method was to be the only practicable way of finding longitude astronomically for the next 1600 years.

In Book I, Chapter IV, entitled 'Carefully observed phenomena should be preferred to those derived from the accounts of travellers' (which was one of Hipparchos's tenets three hundred years before), Ptolemy discusses Hipparchos's lunar-eclipse method of finding longitude:

> . . . and when others coming after him . . . calculating most of their distances, especially those which extend to east or west, from a certain general tradition, not because of any lack of skill or . . . indolence on the part of the writers, but because in their time, the use of exact mathematics had not yet been established; and when in addition to this not many eclipses of the moon have been observed at the same time in different localities as was that eclipse at Arbela which was noted as occurring there at the fifth hour, from which observation it was ascertained how many equinoctial hours [equal as opposed to unequal hours], or by what space of time two places were distant from each other east or west; it is just and right that a geographer about to write a geography should lay as the foundation of his work the phenomena known to him that have been obtained by a more careful observation. . . [2]

The eclipse referred to occurred just before the Battle of Arbela in 330 BC, recorded at both Arbela and Carthage with a supposed difference in local times of three hours. (The actual difference should have been 2 h 15 min.) In fact, this seems to have been the only astronomically determined difference of longitude used by Ptolemy in his *Geography*.[3]

After Ptolemy, more than a thousand years were to pass before there were any significant developments. Then, in the last few decades of the thirteenth century, there were two events of great importance in the story of longitude at sea. The first of these was the appearance in Europe of the mechanical clock, one of the most important inventions of the Middle Ages. As the name implies – the Latin *clocca* means bell – the earliest clocks were primarily for sounding the hours for religious and secular purposes: only later did they become astronomical and navigational instruments.

The second development was the appearance of the sea chart, the earliest reference to which is found in an account of St Louis of France during the Eighth Crusade in 1270. Though most scholars are agreed that there must have been some form of nautical chart in classical times, the sea chart as we know it evolved in Italy and was normally a representation of the Mediterranean drawn in ink on a whole skin of vellum, usually to accompany a *portolano*, or pilotbook (hence such charts came in the nineteenth century to be known as 'portolan charts').[4] The earliest surviving chart is of the Mediterranean, unsigned and undated but, from internal evidence, seeming to have been drawn about 1300. It has no scales of latitude or longitude.[5]

The age of discovery

At the beginning of the fifteenth century trade between Europe and Asia was by way of the Black and Mediterranean Seas, or overland, and was almost entirely in the hands of the Italian maritime states, Venice and Genoa. Cut off from those markets, Portugal looked southward for expansion by sea. Since early in the fourteenth century she had enlisted the services of Genoese pilots to help create a navy. Under the energetic leadership of Prince Henry 'the Navigator' (1394-1460), son of a Portuguese king and an English princess, the Portuguese push to the southward was intensified. Dom Henry was governor of the southern province of Algarve and, from his headquarters at Sagres near Cape St. Vincent, he employed Arab and Jewish mathematicians to instruct his captains in the art of navigation in the Atlantic, an art which was to be very different from that demanded of seamen in the Mediterranean.

Eventually, helped by the revenues of the Order of Christ of which Dom Henry was Grand Master, the Portuguese found a sea-way to the East by sailing southward and eastward round Africa. Cape Bojador, just south of the Canary Islands, was passed in 1434, Cape Verde in 1444, the Cape of Good Hope in 1487, and, in culmination of the Portuguese efforts, Calicut on the Malabar Coast of India was reached by Vasco da Gama in 1498.

These ocean voyages, this sailing into the unknown, gave rise to a new concept in navigation – the use of astronomy to supplement the time-honoured methods of compass, lead-line, and informed estimation of ship's speed. The Portuguese succeeded in devising methods and instruments to find latitude by observation of the Pole

Star and of the Sun. This was adequate as long as the ocean voyages were primarily in a north-south direction, but, once around the Cape of Good Hope, it was east-west distances that mattered – and even more so for Christopher Columbus, taking a departure from the Canaries in September 1492.

Some explorers did try to find their longitude astronomically. To help him do this, Columbus carried with him the *Almanacs* or *Ephemerides* compiled by Johannes Mueller (called Regiomontanus after his birthplace Königsberg) which gave predicted positions of the Sun, Moon and planets for the longitude of the city of Nuremberg for every day from 1474 to 1506,[6] later extended to 1531 by Stöffler of Tübingen.[7] On his second voyage Columbus observed an eclipse of the Moon on 14 September 1494 while at anchor off Hispaniola: the longitude he obtained later turned out to be some 23° too far to the west, about 1½ hours in time.[8] On his fourth voyage, while his ship was aground off Jamaica on 29 February 1504, he used a predicted lunar eclipse, first, to put the fear of God into the natives (as Mark Twain was to do in *A Connecticut Yankee in King Arthur's Court* 350 years later), and then, by his observations, to find his longitude: this time, his error seems to have amounted to more than 2½ hours, again too far to the west. (Whether these errors arose from his own observations of local time, from errors in Regiomontanus's almanac, or whether it was just fudging to prove that he, Columbus, had indeed reached Asia, we may never know.[9]) Amerigo Vespucci's alleged determination of longitude off the coast of South America in 1499, by observing the conjunction of the Moon and the planet Mars, has been shown to be fiction, in which he extrapolated from Columbus's longitudes in the Caribbean.[10]

But the methods claimed by Columbus and Vespucci could be used very infrequently and then generally only in harbour. In the 1490s, therefore, there was a very real need for some solution to the problem of finding longitude *at sea*, a problem which was eventually to lead to the founding of Greenwich Observatory and the establishment of Greenwich time.

Longitude by lunar distance

A possible solution to this problem was indeed forthcoming. In 1514 Johann Werner of Nuremberg (1468-1522) published a new translation of the First Book of Ptolemy's *Geography*.[11] In his commentary on Chapter IV (part of which is quoted above) Werner put forward a new theory for finding longitude which has come to be called the lunar distance method, using a cross-staff, an instrument derived from the Jacob staff invented some two hundred years earlier by the Provençal Hebrew astronomer Levi ben Gerson.[12]

```
                              1 4 0 2
                           Biſſextilis
Aureus numerus  ♉        Quadrageſima       24  Februarii
Cyclus ſolaris       I        Paſca            ∧    Aprilis
Littere domicales   g   f    Rogationes       12    Maii
Inditio             ∧        Aſceſio domini   16    Maii
Intuallũ  ∧  hebdo. 6    dies Pentecoſte     26    Maii
Septuageſima  ♉  Februarii Aduẽtꝰ domĩ     1   Decembris
                         Ecliplis Lunę
                      29    13   36
                           Februarii
                       Dimidia duratio
                            1   26
```

A page from the almanac carried by Columbus, predicting the total eclipse of the Moon on 29 February 1504, which he used to persuade the natives of Jamaica to give his crew food. From the *Ephemerides* of Johannes Mueller, called Regiomontanus, which gave astronomical predictions for the meridian of Nuremberg from 1474 to 1506, from which, in theory, longitude could be found.

```
Saturnus    ab initio āni ad 6 Martii : item a 4 Nouẽbris ad
            exitum anni retrogradus .
Iupiter     a capite anni ad 19 februarii : itẽ a 28 Nouẽbris
            ad calcem anni retrogradus .
Mars        ab initio anni ad 3 Februarii retrogradus .
Venus       ab ineunte anno ad 26 Ianuarii  retrograda .
Mercurius   a 4 Aprilis ad 26 eiuſdem : rurſus a 29 Iulii
            ad 20 Auguſti : et a 22 Nouẽbris ad 18
            Decẽbris retrogradus .
```

Though it took 250 years to become a practical proposition, this method eventually made it possible to measure longitude at sea. Explained more fully in Appendix I, it makes use of the fact that the Moon appears to move comparatively quickly against the background of the stars in the zodiac belt – by approximately its own diameter in one hour. 'Therefore the geographer goes to one of the given places and from there observes, by means of this observational rod [the cross-staff] at any known moment, the distance between the Moon and one of the fixed stars which diverges little or nothing from the ecliptic.'[13] So said Werner, who was talking of the method being used on land. Then, with the aid of astronomical tables for the star's position and an almanac for the predicted position of the Moon, he could find his difference of longitude from whatever place the almanac was based upon. That, however, was only theory. In practice, neither the instruments nor the tables were at that time accurate enough to give a useful result. Furthermore,

The cross-staff in use, both for measuring a lunar distance and for measuring the height of buildings. From the title page of Peter Apian's *Introductio Geographica Petri Apiani in Doctissimus Verneri Annotationes* (Ingolstadt, 1533).

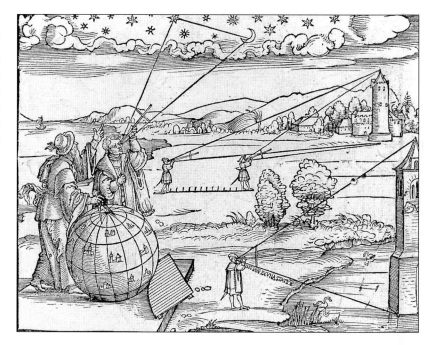

Werner omitted to take into account lunar parallax – the fact that the Moon appears in a different position according to the observer's position on Earth – something which is absolutely fundamental if accuracy is desired.

It was not Werner's own description of the lunar-distance method which made it known to seamen and scholars but that of Peter Apian (1495-1552), whose *Cosmographica* (Ingolstadt, 1524)[14] gave a clearer description as well as a picture. Apian's work was re-edited by Gemma Frisius (1508-55), the Frisian astronomer and mathematician, in 1529[15] and successive editions continued to be published for the rest of the century.

The first description of the method in English was given by Dr William Cuningham of Norwich in his *Cosmographical Glasse* of 1559, a book distinguished by the beauty of both its language and its italic typeface:

> l will shew you, ther are thre thinges required vnto this busines, the Astronomers staffe, also called Iacobes Staffe (the makinge of which you shall finde among th'other instrumentes) the second is the true place of the Mone in the Zodiake, in degrees, & minutes, for the hour you make obseruatiõ, (whiche you may take out of an Ephemerides) and the iij. is the longitude of a fixed sterre, which you may take out of the Table of fixed sterres in my firste boke. These had, you muste take your staffe with the Crosse on it and applye the one ende of the Crosse to the Center of the Mone, and the other vnto the sterre: which thing to do, you shall

The earliest illustration of the lunar-distance method of finding longitude. From the first edition of Peter Apian's *Cosmographica …* (Ingolstadt and Landshut, 1524). In later editions, the observer has grown a beard.

remoue the Crosse vp and downe, vntill the endes of the staffe touch both the center of the mone & also of the sterre. Thys ended, the crosse shall shewe you what the distaunce of the Mone, & starre is in degrees & minutes. Then take the distaunce in degrees, & minuts of the Mone, & fixed sterre, which you had before the obseruation: And substract these ij. distances, th'one out of th'other, the remanet deuide by the portiō that the mone moueth in one hour, And that shall shew you the time, whan as the Mone was ioyned wyth the starre (if the starre be West from her) or when she shall be ioyned with the starre, if it be East from the Mone.[16]

The chronometer method invented

The chronometer method of finding longitude. From W. Cuningham, *The Cosmographical Glasse* (London, 1559), the first treatise in English devoted exclusively to cosmography, and to deal with navigation in relation to astronomy and cosmography.

The first author known to have proposed the employment of a timekeeper for determining longitude at sea is Gemma Frisius in his work *De Principiis Astronomiae Cosmographicae* published in Louvain in 1530. In Chapter 19, 'Concerning a new method of finding longitude', he says:

In our age we have seen some small clocks skilfully produced which are of some use. These, on account of their small size are no burden to a traveller. These will often keep running continuously for up to 24 hours. Indeed if you help, they will keep running as if with perpetual motion. It is therefore with the help of these clocks and by the following methods that longitude is found. In the first place we must take care that before we set out on our journey, we should observe exactly the time at the place from which we are making our journey. Then while we are on our journey we should see to it that our clock never stops. When we have completed a journey of 15 or 20 miles, it may please us to learn the difference of longitude between where we have reached and our place of departure. We must wait until the hand of our clock exactly touches the point of an hour and at the same moment by means of an astrolabe or by means of our globe, we must find out the time of the place at which we now find ourselves. If this time agrees to the minute with the time shown on our watch, it is certain that we are still on the same meridian or in the same longitude, and our journey has been made towards the south. But if it differs by one hour or by a number of minutes, then these should be turned into degrees, or minutes of degrees, by the method I set out in the previous chapter, and in this way the longitude is discovered. In this way I would be able to find the longitude of places, even if I was dragged off unawares across a thousand miles, and even though the distance of my journey was unknown. But then first of all, as always, the latitude must be learnt. I have already explained this before and also that it can be found out by various methods of finding out the tirne. Then indeed it must be a very finely made clock which does not vary with a change of air.[17]

> You shall prepare à parfait clocke artificially made, such as are brought from Flaūders, & we haue thē as excellently without Temple barre, made of our countrymen. Spoud. Do you not meane such, as we vse to weare in the facion of à Tablet? Phi. Yea truely, when as you trauell, you shall set the nedle of youre Diall exactlye on the hour found out by the sonne on the daye, & by some starre in the night: thē traueling withoute intermiſſion, whan as you haue traueled .xx. yea .xl. miles or more (if your next place, whose longitude you desire be so far distant) then marke in your Diall, the houre that it sheweth: after with an Astrolabe, or Quadrant, finde out the hour of the day in that place: & if it agre with the same which your clock sheweth, be assured your place is north or South frō the place you came from , & therfore haue the same lōgitude, & meridiā line. But & the time differ, subtract th'one, out of th'other, & the differēce turn into degrees & minut. of th'equinoctiall as before , then adde or subtract, as in th other .ij. precepts, going before. But now behold the skie is ouer cast with cloudes: wherfore let vs haste to our lodgings, & ende our talke for this presente. Spoud. With a righte good will.

In the 1553 edi[...] [...] first
actual referenc[...]

Therefore it [...] [...]eys, to
use large cl[...] [...]ll measure a whol[...] [...]ck may
be correcte[...]

The first d[...] [...]bove by
Richard E[...]

John Harrison's Timekeepers

Between 1730 and 1760, Harrison built four revolutionary timekeepers in his single-minded pursuit of the longitude prize. Pictured below are the clocks and the dates of their delivery to the Board of Longitude.

H-1 (1737)

H-2 (1741)

H-3 (1759)

H-4 (1760)

As described in
Longitude: The True Story of a Lone Genius Who Solved the Greatest Scientific Problem of His Time by Dava Sobel, published by Fourth Estate Ltd

To order a copy, call
Book Service by Post on 01624 675137
Photographs reproduced by permission of the National Maritime Museum, London.

– all[...] [...]or more than
two[...] [...]on the same
sub[...] [...]etween the
'in[...] [...]roduced on the
op[...]

This gilded brass and steel astronomical clock was made by Casper Buschman II of Augsburg and is dated 1586. The main dial on the top incorporates an astrolabe and shows the position of the Sun and the Moon, and predicts solar and lunar eclipses, but like all clocks of the period was accurate only to around twenty minutes per day.

The longitude prizes

Though theoretical solutions had been found, the practical problem of how to find longitude not only at sea but also on land became the more urgent as oceanic voyages became more frequent and as nations began to depend upon trade with the Indies.

In Spain in 1567 Philip II offered a reward for the solving of the longitude problem at sea. Towards the end of the century Miguel de Cervantes and other Spanish authors began to make fun of the attempts of some 'crazy people' to find position at sea – *el punto fijo*.[21] 'Finding the longitude', along with 'squaring the circle', began to be equated with something which was virtually impossible: Cervantes's Mathematician in *The Dog's Dialogue* says:

> I have spent twenty-two years searching for the fixed point *[punto fijo]* and here it leaves me, and there I have it, and when it seems I really have it and it cannot possibly escape me, then, when I am not looking, I find myself so far away again that I am astonished. The same thing happens with squaring the circle, where I have arrived so near to the point of discovering it that I do not know and cannot imagine how it is that I have not got it in my pocket . . .[22]

In 1598 Philip III offered 6,000 ducats as perpetual income, plus 2,000 as income for life, and 1,000 for help in regard to expenses as a reward to anyone who could 'discover the longitude'. The whole prize was never won but considerable sums were disbursed on account (so to speak) to encourage possible inventors. Gould quotes seven instances of grants between 1607 and 1626, most of which were concerned with compasses and magnetism.[23] About the same date, the States General of Holland offered 30,000 florins.[24] Portugal and Venice are also said to have offered rewards.[25]

The most famous person to apply for the Spanish prize was probably Galileo Galilei (1564-1642), the Italian astronomer. One of his first celestial discoveries with the newly invented telescope was that Jupiter had four moons in orbit (twelve more have been discovered since). With orbital periods varying between 1¾ and 17 days, they appear and disappear (as seen from Earth) as they pass behind Jupiter itself and their reflected sunlight is eclipsed each time one of them passes into Jupiter's shadow. The eclipses always occur at precisely the same moment for an observer anywhere on Earth, and the occultations virtually so. Galileo realized that here he had a perfect celestial timekeeper which, if the eclipses could be accurately predicted, could be used for longitude-finding exactly as Hipparchos had suggested using the eclipses of our own Moon – but with the added advantage that the eclipses of the former occur much more frequently, once or twice a night.

But there were disadvantages: firstly, one had to use a telescope, which it was thought would be difficult (later, it proved impossible) to use at sea; secondly, the eclipses were not quite instantaneous. However, Galileo, confident that these difficulties could be overcome, set about studying the motions of the satellites and drawing up tables of the predicted times of eclipses. In 1616 he submitted this method for the Spanish longitude award. The Spaniards were unimpressed and, after a protracted correspondence, Galileo seems to have given up the idea of selling it to Spain in 1632. In 1636 he tried Holland, saying he had spent twenty-four years perfecting his tables. Unlike Spain, the States General *were* impressed, but negotiations were difficult. Galileo was at that stage virtually under house arrest at Arcetri near Florence, being closely supervised by the Inquisition who, it is said, refused to allow him to accept the gold chain awarded him by the Dutch Government. His death in 1642 ended the negotiations.

Seventeenth-century developments

Galileo made a second contribution to the longitude story – and a significant contribution to horology – by his study of the pendulum as a method of regulating timekeepers. The first recorded use of a weight-driven clock for astronomical purposes seems to have been in 1484 when Bernard Walther, a pupil of Regiomontanus, used such a clock to measure the interval between the rising of the planet Mercury and the moment of sunrise.[26] Clocks were also part of the equipment of the great Danish astronomer Tycho Brahe (1546-1601), whose catalogue of 777 stars – enlarged to some 1,005 and published by Galileo's contemporary Johann Kepler (1571-1630) in his Rudolphine Tables of 1627[27] – was still the best available when the Royal Observatory was founded in 1675. Tycho purchased and tried four clocks between 1577 and 1581 before concluding that the inherent defects of sixteenth-century clockwork were too great for most astronomical purposes.

Galileo first drew attention to the value of the pendulum as a controller for clocks in 1637, but it was not until the last years of his life, 1641-2, that he developed a practical mechanism. Because it is controlled by gravity, the pendulum has a natural, regular motion. It is isochronous, that is, it oscillates in equal spaces of time almost irrespective of the arc of swing or weight of the bob. The time of swing, however, does vary in relation to the pendulum's length, and it is therefore easily adjusted.

Galileo produced designs for a rudimentary pendulum clock with an escapement to keep the pendulum in motion. However, it is believed that no actual clock was ever made to his design, so the invention of the pendulum clock – as opposed to the discovery of the isochronism of the pendulum – is generally attributed to the Dutch mathematician, astronomer and horologist, Christiaan

Pocket globes were invented largely as teaching tools. The outer globe serves as a case and has a map of the heavens. The inner globe is of the Earth and shows the Greenwich meridian running south from the Arctic Circle. By George Adams of London, c. 1755.

Huygens (1629-96). The invention of the pendulum clock in 1657 and of the balance spring for watches in the 1670s – in which Huygens also had a hand – were two of the most fundamental developments in the science of horology.

Before returning to Huygens and his timekeepers, we should mention a theoretical development in the search for longitude. In 1634 Jean-Baptiste Morin, doctor of medicine and professor of mathematics at the Collège Royal in Paris, announced that he had discovered the secret of longitude. Cardinal Richelieu thereupon set up a commission of an admiral and five scholars to examine this claim. An astrologer and believer in the Ptolemaic Earth-centred universe, Morin distrusted clocks and is reported to have said that he did not know whether the Devil would succeed in making a longitude timekeeper but that it was folly for a man to try.[28] Basically, he proposed the lunar-distance method already outlined, but with the added refinement that he took account of lunar parallax, a very important development (see p. 22 above). Morin's method was geometrically sound but the commission considered he had not 'found the longitude' because the imperfections of the tables of the stars and Moon – which was no fault of Morin's – meant that it could not be used in practice. Morin applied to Holland for a reward without success. Eventually, he was granted a pension of 2,000 livres in 1645 by Richelieu's successor, Cardinal Mazarin.

Having invented the pendulum clock in 1657, Huygens turned his attention to the longitude problem, convinced that the horological approach – to produce a marine timekeeper that would keep accurate and regular time for months in any climate, regardless of the motion of the ship – would soon make it possible to discover the longitude. He produced various marine timekeepers which were tried at sea between 1662 and 1687. In 1668 one of his timekeepers, which had kept going during both gales and a sea battle, gave a difference of longitude between Toulon and Crete as 20° 30′ as against the true value of 19° 13′, an error of only 100 km or so.[29] His early timekeepers were controlled by pendulums but, in anything but a flat calm, their going was most erratic. So, in 1674, he abandoned that design and proposed to control his marine timekeepers with balance and balance spring. Huygens made a brave effort but, as we shall see, it was to be a hundred years before a satisfactory marine timekeeper could be produced.

When his longitude proposals had been turned down in the 1630s, Morin had suggested that what was needed to make them practicable was an observatory to provide the necessary data. This was achieved with Louis XIV's foundation of Paris Observatory in 1667, resulting from the foundation of the Académie Royale des Sciences the previous year, largely at the instigation of Louis's

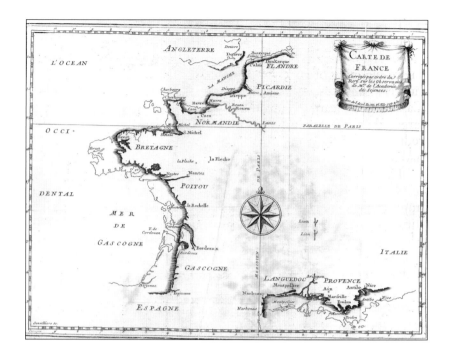

Finance Minister Jean-Baptiste Colbert, who was determined to make France pre-eminent in science – and at sea.

The British equivalent of the French Académie des Sciences, the Royal Society of London for Improving Natural Knowledge, had been founded by Charles II four years earlier, in 1662. 'Finding the longitude' was only one of the many subjects in natural philosophy which engaged the attention of this newly formed learned society. The following is verse 26 of a poem entitled 'In Praise of the Choyce Company of Philosophers and witts who meete on Wednesdays weekly at Gresham Colledge':

26.   The Colledge will the whole world measure,
    Which most impossible conclude,
  And Navigators make a pleasure
    By finding out the longitude.
  Every Tarpalling[30] shall then with ease
    Sayle any ships to th'Antipodes.

From internal evidence this poem seems to have been written about 1661, which was a year before the King granted his Royal Charter. Of twenty-eight verses altogether, its previous twenty-five describe the various other projects considered by the embryo society – Wren's Moon globe, Boyle's air pump, magnetic experiments with iron filings, how moths eat cloth, Evelyn's diving bell, graving and etching, remarks on smoke pollution, and Wilkins's Universal Character (an artificial language and script).[31]

Two matters concerning the French Académie at this time should be mentioned before we embark on the story of Greenwich time. One of the first projects undertaken by the new academy was the re-mapping of France, with longitudes found by observation of Jupiter's satellites, as first suggested by Galileo. The re-survey by Cassini and Picard proved that the old maps showed France to be larger than it actually was. The King, displeased by the apparent reduction of his territory, is reported to have said that his surveyors had lost him more land than his armies had gained.

The second matter concerned a German called André Reusner of Neystett who, in 1668, approached Louis XIV with a reputedly impeccable solution to the longitude problem – an 'odometer', a form of ship's log capable of measuring not only speed through the water but, according to its inventor, speed over the ground as well. 'He addressed himself to the king and obtained a letter of intent by which His Majesty, having undertaken to finance a discovery from which all nations would profit, promised to pay the inventor 60,000 livres in one lump sum, and to grant him the right to draw 4 sous for every ton of capacity of all vessels availing themselves of the discovery. His Majesty engaged himself to guarantee this right up to the sum of 8,000 livres per annum, reserving for himself nothing more than the option to withdraw this right on payment of 100,000 livres. One sole condition attached to these magnificent promises: that the inventor demonstrate the effectiveness of his invention before M. Colbert, M. du Quesne, Lieutenant General in His Majesty's naval forces, and MM. Hughuens, Carcavy, Roberval, Picard, & Auzout of the Academy of Sciences.' The Commissioners duly met and concluded that, although it was ingeniously devised, it was not capable of finding longitude with any certainty. 'The German was required to answer in writing all the problems raised by the Academy, and he did so, but not even the 160,000 livres already guaranteed could inspire him to find answers appropriate to annul these objections.'[32] So Reusner never got his money. Nevertheless, the story shows how much importance the French attached to finding a solution to the problem of discovering longitude at sea. As the same account says: 'A solution in this matter of longitude would be of the greatest possible utility to the public as well as its Author; for great rewards are promised whomsoever shall first resolve this Problem. Seeing indeed that so many men have been at pains to solve the Quadrature of the Circle, though that could bring them naught but glory, they cannot have been negligent in finding out Longitude, which would bring no less glory but incomparably more profit.'[33] This search did indeed occupy the minds of many men, for many years to come.

PROSPECTUS INTRA CAMERAM STELLATAM

TWO

Greenwich time for astronomers

1675 - 1720

THE story of Greenwich time started in England in 1674, when Charles II had been on the throne for fourteen years. England was for the time being at peace, following her somewhat unnatural (and domestically very unpopular) alliance with France during the Third Dutch War, an alliance soon terminated after the Dutch Admiral De Ruyter's naval successes had heightened the unpopularity of the war in England.

Three people require introduction, all of whom had a great influence – the first unwittingly – upon the history of astronomy and science generally and upon this story in particular. The first of these is Louise de Kéroualle (1649-1734), a Breton lady who, if she never monopolized King Charles II's affections, at least secured a greater share of them than any other woman. Daughter of Guillaume Penancoët, Sieur de Kéroualle in Brittany, Louise was maid of honour to the Duchess of Orléans, Charles' youngest and favourite sister Henriette-Anne ('Minette'), who married Louis XIV's brother Philippe in 1661 and thereby became the second lady of France.

After the Duchess's untimely death from peritonitis in 1670, Louise was brought to England by the Duke of Buckingham – some say was sent by Louis XIV as a spy – to be named maid-of-honour to Queen Catharine. The King had grown tired of Lady Castlemaine (pensioned off and created Duchess of Cleveland in 1670), so Louise, with her dark good looks and baby face, soon established herself as 'Old Rowley's' mistress, bearing him a son, Charles Lennox, later Duke of Richmond, on 29 July 1672. After naturalization Louise was created Duchess of Portsmouth in 1673 and was henceforward always regarded as the King's principal mistress.

Opposite. The Great Room (now the Octagon Room) in the Royal Observatory, which has not changed much since it was built in 1675. The etching, of about 1676, was by Francis Place, after Robert Thacker.

Louise de Kéroualle, Duchess of Portsmouth. Detail from a portrait by Sir Peter Lely.

33

Effigies Ionæ ꝰꝏꝏꝏMoore *Matheseos*
Professoris Ætat: suæ · 45 · An: Dñi:1660

Sir Jonas Moore, Surveyor-General of the Ordnance, who in 1675 persuaded King Charles II that he should found a Royal Observatory at Greenwich to provide astronomical data for finding longitude at sea, From an engraving in his *A New System of Mathematics* (1681).

The second is Jonas Moore (1627-79), born at Whitelee in Pendle Forest, Lancashire. He resolved early to devote himself to mathematics. During the Civil War he had the use of the library of the antiquarian Christopher Towneley, the patron of many young north-country natural philosophers such as William Gascoigne, Jeremiah Horrocks, Jeremiah Shakerley, and Christopher's own nephew, Richard Towneley. Publishing his first work, a mathematical textbook, in 1647, Moore came to London soon after, setting up as a mathematics teacher. He found all too few pupils during those troubled times, but was fortunate enough to be appointed surveyor in the work of draining the Great Level of the Fens from 1649, a post in which he made his reputation, being kept busy thereafter with various other surveys throughout the Commonwealth period. On the Restoration he re-published his *Arithmetick*, with a dedication to the Duke of York. In 1662 Tangier came to the British Crown as part of the dowry of Queen Catherine of Braganza, and Moore was sent there in 1663 to survey and report on the fortifications.

In 1669 he was knighted and appointed Surveyor-General of the Ordnance, henceforward residing in the Tower of London, where 'he enjoyed high royal favour, which he turned to account for rescuing scientific merit from neglect',[1] one of the manifestations of which was his becoming, in 1670, the patron of the 24-year-old astronomer John Flamsteed. Aubrey called Moore 'one of the most accomplished gentlemen of his time: a good mathematician, and a good fellow', adding, 'Sciatica: he cured it by boiling his buttock'.[2] In 1673 Moore was, with Samuel Pepys, one of those responsible for the foundation of the 'Royal Mathematical School within Christ's Hospital' to provide training in navigation for boys for the King's service at sea.[3]

The third character is John Flamsteed (1646-1719), born near Derby, the son of a rich maltster. At the age of 14 he was afflicted by 'a fit of sickness that was followed with a consumption, and other distempers',[4] which caused him to quit the Free School in Derby two years later and left him with a delicate constitution, which was to plague him for the rest of his life. It also meant that he could not immediately go up to university but had to pursue his studies at home, principally in the sciences of mathematics and astronomy. By 1669 he was corresponding with Fellows of the Royal Society in London, which he visited at Easter 1670. There he met Sir Jonas Moore who, recognizing his talents, became his patron and presented him with astronomical instruments – telescope glasses and tubes, and a micrometer – with which to pursue his researches in Derby. On the way home he visited Cambridge, entering his name for Jesus College and meeting, among others, Isaac Newton, with whom he would later have so many disagreements. In 1672 he visited Richard Towneley at Towneley Hall, near Burnley in Lancashire, to study the observations of the brilliant north-country astronomers – William Gascoigne, Jeremiah Horrocks and William Crabtree – all of whom had

JOHANNES FLAMSTEEDIUS Derbiensis Astronomiæ Professor Regius. Anno Ætatis 74 Obijt Decem. 31 1719

died in or before the Civil War, and whose papers had been preserved by Richard Towneley's antiquarian uncle, Christopher.

Thanks to the interest and support of Sir Jonas Moore, the King issued a warrant to Cambridge University on 14 May 1674 'to grant an MA degree to John Flamsteed, late of Jesus College, who has spent many years in the study of the liberal arts and sciences, and especially of astronomy, in which he has already made such useful observations as are well esteemed by persons eminently learned in that science'.[5]

John Flamsteed was appointed first Astronomer Royal aged 29 and died in the post at Greenwich over forty years later. Engraved by George Vertue as frontispiece to his *Historia Coelestis* (1725), after a portrait of 1712 by Thomas Gibson.

The foundation of Greenwich Observatory

In the autumn of 1674 the Royal Society began to make plans for setting up an observatory in King James I's old College at Chelsea (demolished in 1682 to make way for Wren's Royal Hospital) which had been presented to the Society in 1667. Sir Jonas Moore offered to pay for all expenses connected with this observatory and proposed that the 28-year-old Flamsteed, who was preparing to take Holy Orders the following Easter, should be observer.

However, while these plans were being thought about there occurred an event which was to have a profound effect upon our story. In Flamsteed's words, 'An accident happened that hastened, if it did not occasion, the building of the [Greenwich] Observatory. A Frenchman, that called himself Le Sieur de St. Pierre, having some small skill in astronomy, and made [sic] an interest with a French lady, then in favour at Court, proposed no less than the discovery of the Longitude.'[6] The lady then in favour at Court was none other than the Duchess of Portsmouth. We still know almost nothing about St. Pierre but, from his name and title, it seems quite likely that he came from Brittany and may well have had previous links with the Duchess; but research has so far failed to identify him.[7] Be that as it may, the King was persuaded to issue a warrant on 15 December 1674 appointing a Royal Commission to examine the Sieur's proposals:

> that he [St. Pierre] hath found out the true knowledge of the Longitude, and desires to be put on Tryall thereof; Wee having taken the Same into Our consideration, and being willing to give all fitting encouragement to an Undertaking soe beneficiall to the Publick
>
> . . . hereby doe constitute and appoint you [the Commissioners], or any four of you, to meet together . . .
>
> And You are to call to your assistance such Persons, as You shall think fit: And Our pleasure is that when you have had sufficient Tryalls of his Skill in this matter of finding out the true Longitude from such observations, as You shall have made and given him, that you make Report thereof together with your opinions thereupon, how farre it may be Practicable and usefull to the Publick.'[8]

The Commissioners were a distinguished group:

Lord Brouncker, President of the Royal Society, and Controller of the Navy.

Seth Ward, Bishop of Salisbury (Sarum), Savilian Professor of Astronomy at Oxford 1649-61.

Sir Samuel Morland, mathematician, inventor and Gentleman of the Bedchamber to Charles II.

Sir Christopher Wren, the King's Surveyor-General, Gresham Professor of Geometry in London 1657-61, and Savilian Professor of Astronomy at Oxford 1661-73.

Col. Silius Titus, Gentleman of the Bedchamber.

Dr. John Pell, mathematician, principally known for having invented the sign ÷ for division.

Robert Hooke, MA, surveyor, curator to the Royal Society, and Gresham Professor of Geometry.

All of these except Wren were already members of a Committee

Opposite.
John Flamsteed and his assistant Thomas Weston, sketched from life in 1710 by Sir James Thornhill, and painted in the south-east corner of the ceiling of the Painted Hall, Greenwich Hospital, today's Royal Naval College.

John Flamsteed sketched in an observation book by his assistant Abraham Sharp, 1684.

appointed by the King in the previous year to investigate the merits of proposals by one Henry Bond for finding longitude at sea by measuring the dip of the magnetic needle.[9] As for Wren, before he turned to architecture he had been an astronomer, so his appointment to the new Committee was most apposite.

On 2 February 1675 John Flamsteed arrived in London to stay in the Tower with Sir Jonas Moore. A few days later Moore contrived to have Colonel Titus take Flamsteed to the King to inform His Majesty of the very important results being achieved by Jean Picard and his colleagues of the Académie des Sciences in their great survey of France. In order to determine longitude Picard was observing the eclipses of the moons of the planet Jupiter, a method very successful ashore, but one which was to prove impracticable at sea because the ship's motion made accurate observations impossible. In the light of subsequent events it seems not improbable that Flamsteed also discussed with the King the problem of longitude at sea and the urgent need for an observatory in England to provide the basic data. True, the Royal Society had plans for such an observatory in Chelsea, in which Flamsteed was himself involved. But did Flamsteed remind the King that Picard had the resources of a *royal* observatory behind him, an observatory – Paris Observatory – founded by the French King eight years before?

Meanwhile, St. Pierre – and his patron the Duchess – had been getting impatient. On 31 January 1675 the Secretary of State, Sir Joseph Williamson, told Morland that the Committee must set to work immediately. The next day Morland sent a letter to Pell (who was acting as secretary) by the hand of St. Pierre himself. Eventually, four out of his seven commissioners met at Col. Titus's house on Friday 12 February, Morland, Pell, Titus and Hooke providing the quorum of four demanded. At Sir Jonas's instigation (though not himself a member), Flamsteed was admitted as an official Assistant to the Committee.

> He [St. Pierre] pretended no less than the absolute discovery of the longitude from easy celestial observations, and demanded the heights of two stars, and on which side of the meridian they were, with heights of the Moon's two limbs [upper and lowerl with the pole's height to be given in minutes [i.e. he required the latitude of the observer], as also the year and day of observations, whence he undertook to show under what meridian these observations were made.[10]

So wrote Flamsteed in 1682. Though not specifically stated, it is almost certain that St. Pierre was hoping for a large reward for finding the solution to this vital problem.

At the St. Pierre Committee's first meeting in London on Friday 12 February, Flamsteed undertook to supply the observations called for and these were duly delivered to Pell the following Wednesday.

Pell passed them on to St. Pierre two days later:

> On Friday, Febr. 19. I delivered those two observations to that
> Frenchman in the presence of Mr. Payen, a Lorrain Gentleman belong-
> ing to Secretary Coventry. This Mr. de St. Pierre disliked that they were
> for the years 1672 & 1673. He would have had them for the year 1675. I
> told him that if he could obtain that the King should send observers into
> the Indies, he must tarry till they could come back hither again. He said:
> these were but Calculations, & he could doe nothing with them. I
> answered, that if they were but calculations we could easily have calculat-
> ed for any night of this February, &c.[11]

Meanwhile Flamsteed had written down his comments on St.
Pierre's proposals. Yes, he said, the Frenchman's method might
work in theory, though better methods were known. St. Pierre proposed to use vertical angles to measure the Moon's longitude
whereas (as can be seen in Appendix I) the Moon's movement
among the stars is basically horizontal. Furthermore, such vertical
measurements were made the more uncertain by atmospheric refraction (which varied according to the weather) and because of
imperfect knowledge of the Moon's parallax. Flamsteed's 'better
method' was, of course, that of lunar distances (see Appendix I),
where the principal observation is more horizontal than vertical.

But, wrote Flamsteed, whatever method was used, there was a
fundamental limitation in that, with the then state of knowledge,
the basic information simply was not available to the degree of
accuracy needed. For any lunar method one must (a) know where
the so-called fixed stars are, relative to the Sun's annual path (the
ecliptic), and (b) be able to predict, perhaps years ahead, where the
Moon will be relative to the stars at the moment the navigator takes
his sight. From his own observations Flamsteed had proved that the
best available star catalogue, that of Tycho Brahe, could be in error
by ten minutes or more, while current tables could err by as much
as twenty minutes. All of which meant that any longitude found
could be in error by several hundreds of miles. To provide the necessary data, said Flamsteed, would demand years of observations
with large instruments fitted with telescopic sights. And so it
proved: the search for the data demanded by Flamsteed was to
occupy astronomers the world over for the next 150 years.

On Wednesday 3 March the Committee – Bishop Sarum,
Morland, Titus, Pell and Hooke – met again at Titus's house, this
time to consider Henry Bond's longitude proposals and to agree on
a report to the King. At that meeting it was decided that the report
should be presented to the King by Sarum, Morland, Titus and Pell
at the Privy Gallery, Whitehall at 8 o'clock the following morning.
Hooke was furious at being excluded: 'Titus a dog,' he wrote in his
diary for 3 March, 'I should have been at the King's next morning

The Royal Observatory at Greenwich from Croom's Hill, about 1680, soon after its completion, with Inigo Jones's Queen's House (*centre left, at bottom of hill*), now part of the National Maritime Museum, and the King's House (*far left, near the river*), now part of the Royal Naval College. The 80 ft mast in the Observatory garden supported a refracting telescope of 60 ft focal length. It was never a success and was removed in 1690 or soon after. From an oil painting by an unidentified artist.

with the report.' On 7 March he added: 'Mr Hill told me of Dog Titus his abuse. Query Lord Sarum.' Unfortunately we do not know the other side of the story. In the event, only the Bishop and Titus entered the King's Bedchamber to present the Bond report. But apparently they presented also Flamsteed's report on the St. Pierre proposals:

> When Charles II, King of England, was informed . . . [that the information for finding longitude by lunar observations was not available], he said the work must be carried out in royal fashion. He certainly did not

want his ship-owners and sailors to be deprived of any help the Heavens could supply, whereby navigation could be made safer. [12]

And in another account, Flamsteed goes on:

the ingenious gentlemen ... therefore readily joined with Sir Jonas Moore to move the King that an observatory might be built and furnished with convenient instruments for making such observations as were necessary for correcting the places of the Fixed Stars, the Luminaries and Planets, in order to the Discovery of the Longitude which was not to be otherways expected, and myself to be employed in

> it, with a salary for my support in the work, which His Majesty was gra-
> ciously pleased to grant . . .[13]

So, without waiting for any of the Frenchman's results, the King signed a royal warrant that very day, Thursday 4 March 1675, appointing John Flamsteed his 'astronomical observator', enjoining him 'forthwith to apply himself with the most exact Care and Diligence to the rectifying the Tables of the Motions of the Heavens, and the places of the fixed Stars, so as to find out the so much desired Longitude of Places for perfecting the art of Navigation';[14] in other words, to provide the observational data so that lunar distances could be predicted. The warrant was addressed to the Gentlemen of the Ordnance, one of whom was Sir Jonas Moore, Surveyor-General of the Ordnance. The Office of Ordnance was to pay Flamsteed the far-from-princely salary of £100 per annum, effective from the Michaelmas before.

Sir Jonas brought the news to Flamsteed, who was disappointed with the meagre stipend. Writing in the third person in 1710 he says ruefully: 'A larger salary was designed him at first; but, on his taking orders, it was sunk to this.'[15] He received Holy Orders at the hands of Bishop Gunning at Ely Palace in London at Easter 1675.

> The next thing to be thought of was a place to fix in. Several were pro-
> posed as Hyde Park and Chelsea College. I went to view the ruins of this
> latter and judged it might serve the turn: and better because it was near
> the Court. Sir Jonas rather inclined to Hyde Park, but Sir Christopher
> Wren mentioning Greenwich Hill, it was resolved on . . .[16]

So wrote Flamsteed in 1707. All three sites were Crown property, an important consideration in the current state of royal finances. There were already plans for an observatory at Chelsea. Of the site in Hyde Park we know nothing. But Wren, the King's Surveyor-General, who had relinquished the chair of Astronomy at Oxford only two years earlier, chose Greenwich Castle – on high ground overlooking the River Thames, in the centre of a royal park, away from the smoke of London but accessible by road and river.

The choice must have been quickly made. On the Saturday, only two days after the King signed the warrant, Hooke says in his diary: 'At Sir J. Moores . . . He procured a patent for Flamsteed of £100 per annum and an observatory in Greenwich Park.' By Sunday the Frenchman had heard the news too. He came hotfoot to see Pell without an interpreter. Speaking in Latin, he asked two questions about the observations Flamsteed had provided, questions which caused Flamsteed to declare subsequently that St. Pierre could not possibly know what he was talking about. St. Pierre then demanded further observations because he declared that the previous ones were *fictae et absurdae* – fictitious and ridiculous.[17]

Pell did nothing. However, St. Pierre – and presumably the Duchess also – continued to importune the King, claiming that Flamsteed's observations had been invented – which, in the strictest sense, was true.[18] On 23 April, seven weeks after Flamsteed's appointment, Pell received a letter from Secretary Williamson:

> Whitehall
> 23 Apr. 1675
>
> Sir
>
> His Majesty is so daily importuned by Mons. St. Pierre the French Longitude Man, that he commands me to signify to you, that absolutely to have a finall answer, that is that you forthwith give him such Data he pretends are necessary to the work in hand, and that we may either have the great service he pretends to, or at least a quiet from his further importuning etc. You must please to set yourself to this forthwith & not intermit a day till it be finished, and an account returned to His Majesty.[19]

Secure in his position and already signing himself *As: Re*: (Royal Astronomer), Flamsteed immediately wrote two more scathing reports, one in Latin for St. Pierre, one in English for Pell and the King. He said, in effect, that the Frenchman did not know what he was talking about and in any case had probably borrowed what little good there was in his method from Longomontanus and Morin, two long-dead astronomers who had submitted solutions to the problem many years before.

> 'He had no way to come off but by pretending that the observations were feigned: I showed him that they were not, yet had they been so, they might have served for his purpose in some cases; that he had only betrayed his own ignorance; and that we knew better methods. Upon which, he huffed a little, and disappeared; since which time we have heard no further of him.'[20]

Poor St. Pierre! Though he failed to get his hoped-for reward – and though his identity is still a mystery – he has nevertheless secured for himself a place in the history of astronomy by acting unwittingly as a catalyst in the foundation of the Royal Observatory at Greenwich.

The building of Greenwich Observatory

Though the choice of site at Greenwich had been quickly made, the actual arrangements for the building took rather longer. While Flamsteed was waiting he took up residence with Sir Jonas in the Tower of London, using the north-east turret of the White Tower as a temporary observatory. A nice story, probably apocryphal, is told to the tourists by the yeomen warders at the Tower today, that Flamsteed complained of the nuisance caused by the many ravens then in the Tower, presumably because they sat on and fouled his telescopes. The King was about to give orders for all the ravens to be disposed of when he was informed of a tradition which said that, when the ravens left the Tower, the Tower would fall, and probably the Throne also. Mindful of recent constitutional events, Charles is said to have so modified his orders that a limited number of ravens should always be kept, which is the reason why one of the yeomen warders is still appointed 'raven master'.

At last, on 22 June 1675, more than three months after first discussions, the King sent Sir Thomas Chicheley, Master-General of the Ordnance, a warrant authorizing the building:

> Whereas, in order to the finding out of the longitude of places for perfecting navigation and astronomy, we have resolved to build a small observatory within Our Park at Greenwich upon the highest ground at or near the Place where the Castle stood, with lodging rooms for Our Astronomical Observator and Assistant. Our Will and Pleasure is that according to such plot and design as shall be given you by Our Trusty and well-beloved Sir Christopher Wren Knight, Our Surveyor General, of the place and scite of the said Observatory, you cause the same to be fenced in, built and finished . . .[21]

The castle referred to was on the site of a tower built by Humphrey, Duke of Gloucester (brother to Henry V), soon after the park was enclosed in 1437. The tower was rebuilt in 1526 and used as a guest house and hunting lodge by Henry VIII.

In Queen Elizabeth's time the tower was sometimes called Mirefleur and came to be known as Greenwich Castle. When in 1579 the Queen heard that her favourite Robert Dudley, Earl of Leicester, had married Lettice Knollys the previous year, she 'thereupon grew into such a passion, that she commanded Leicester not to stir out of the Castle of Greenwich, and intended to have him committed to the Tower of London . . .'[22] In the event, Leicester was not sent to the Tower and was soon back in the Queen's favour. During the Civil War the castle was garrisoned by Parliament. The date it was demolished is not known: it was certainly standing in 1662 when it is seen in a sketch on Jonas Moore's map of the Thames,[23] while the King's warrant quoted above implies that it was down by June 1675, which perhaps accounted for the delay in setting the building of the observatory in train.

In the latter part of the warrant the King directed Chicheley to give orders to the Treasurer of the Ordnance, Sir George Wharton: '. . . for the paying for such materials and workmen as shall be used and employed therein, and of such moneys as shall come into his hands for old and decayed powder which hath or shall be sold by Our Order of the first of January last; provided that the whole sum to be expended and paid shall not exceed £500 . . .'[24] The same day that the warrant was signed, 22 June, Robert Hooke was asked by

Wren to 'direct' the building of the observatory. 'He promised money,' added Hooke hopefully. On 30 June Hooke visited the site with Flamsteed and Edmond Halley (who was to become the second Astronomer Royal). On 28 July Hooke went down to Greenwich with Sir Jonas Moore and party and 'set out' the observatory, presumably to the designs of Wren. In July Flamsteed moved from the Tower to apartments in the Queen's House at the bottom of what is known today as Observatory Hill, in order to be on hand to superintend the building. He received an imprest for the first £100 for the building on 27 July and himself laid the foundation stone on 10 August.

Economy was the order of the day. To save money, the new building was raised on the foundations of the old castle, resulting in its walls being 13½° away from true north – which Flamsteed thought was a pity. Lead, wood and iron came from the demolished Coldharbour gatehouse at the Tower; bricks came from Tilbury Fort; two copper balls and two great round shot, presumably for the roof-turrets and gateposts, came from Ordnance stores; spars for the long telescopes came from the Navy Board.[25] And labour and other materials were paid for by the sale of 690 barrels of old and decayed gunpowder from Portsmouth and the Tower to Mr Polycarpus Wharton at 40s. a barrel – who presumably made it serviceable again and re-sold it to the Ordnance at £4 a barrel.[26] The whole cost £520. 9s. 1d., just £20. 9s. 1d. overspent.

The roof was on by Christmas. On 29 May 1676 – the King's birthday – Flamsteed moved some of his apparatus from the Queen's House to the Great Room of the new observatory in preparation for a partial solar eclipse on 1 June (civil date) which the King said he wished to observe from his new Royal Observatory. In the event, the King did not come but was represented by Lord Brouncker, President of the Royal Society, one of the original St. Pierre committee.

Flamsteed and his staff of two moved into the house on 10 July. On 15 September Sir Jonas Moore, Sir John and Lady Hoskins, and Robert Hooke went to Greenwich where they found Flamsteed and the 19-year-old Halley – then friends, though they were to quarrel bitterly in later years – putting the finishing touches to the new instruments. The following day, 16 September 1676, the first observations with the great sextant were recorded: the work of the Greenwich Observatory had begun.

The Great Clocks

We come at last to Greenwich time proper. As we have seen, two of the ingredients in any solution to the problem of finding longitudes at sea by the lunar-distance method were accurate star catalogues and accurate tables of the motion of the Moon: it was specifically to provide these that Greenwich Observatory had been founded. A third ingredient was the provision of an accurate instrument with which to measure the angular distance between the Moon and star or Sun, and to measure the altitudes of these heavenly bodies above the horizon: this will be discussed briefly in the next chapter.

But there was a fourth ingredient. Any astronomical method of finding differences of longitude must make the assumption that, for practical purposes, the Earth rotates on its axis at a constant speed. This had always been *assumed* by the Copernicans but it had never been *proved*. By good fortune, at the very time the Royal Observatory was founded, the means of proving this had just come to hand – the pendulum clock, invented by Huygens as recently as 1657. Moore and Flamsteed determined that one of the first tasks of the new observatory should be to conduct experiments to obtain this proof, using clocks more accurate than any that had ever before been made.

Accordingly, late in 1675, Sir Jonas ordered two clocks for the new observatory from Thomas Tompion (1638-1713), the leading clockmaker working in London. These Great Clocks, as they came to be called, had several very special features designed to make them the most accurate clocks in the world:

(*a*) pendulums 13 feet long (3.96 metres) hung *above* the movements, beating every two seconds rather than the more usual one second of the 'royal' pendulum of 39 inches (1 metre) and hung below the movements, recently introduced and named for the King himself (see p. 162);

(*b*) pin-wheel escapements, apparently dead-beat, based partly upon the ideas of Richard Towneley, friend of both Moore and Flamsteed;

(*c*) very heavy driving weights, so hung that the clock had to be wound only once a year. [27]

The two clocks were most unusual and may even have influenced Christopher Wren in the design of the Great Room, which is the central feature of the building itself, today called Flamsteed House. The room was capable of accommodating the two clocks, with their long pendulums and movements behind the wainscot. This can be seen in the etching on p. 32, where the dials (with the pendulum bobs showing through the little windows above) are to the left of the door. In that same picture can also be seen the other reason for the lofty and dignified room – the tall windows for use with the long telescopes of the day.

When some six years later Bishop Fell was proposing that the new Tom Tower in Christ Church, Oxford, should be fitted up as an observatory, Wren tried to discourage him, by saying that in these days astronomical work was better carried out at ground level. 'Wee built indeed an Observatory at Greenwich not unlike what your Tower will prove, it was for the Observator's habitation & a little for Pompe; It is the instruments in the Court after the manner I have described [mural quadrant on a meridian wall, long telescope suspended from a mast, sextant for angular distances] which are used, the roome keeps the Clocks & the Instruments that are layed by.'[28]

In a letter to Richard Towneley on 22 January 1676 Flamsteed says: 'Our great room being 20 foot high will be capable of a long pendulum to be hung above the clock.'[29] On 6 July the same year he reports: 'Wee shall have a payre of Watch clocks down here to morrow with pendulums of 13 foot and pallets partly after your manner, with which I hope wee may trye those experiments much more accurately then with the second pendulum which I find when the wheels are a little swarft with dust goes much swifter than it ought, sometimes almost a ¼ of a minute per diem.'[30]

The dials of these clocks (which still survive) all have the following inscription: 'Motus Annuus [Year Movement]. Sr Jonas Moore Caused this Movement with great Care to be Made A° 1676 by Tho. Tompion.' Over two months were to pass before they were in going order, but on 24 September 1676 Flamsteed reported: '3^h p.m. Both movements set together.' The next day he wrote, 'Sept. 25. 8.55.00 by the spring hung pend[ulum]: set mine [J.F.'s own clock] with it. 8.54.20 *circiter* by the pivot pendulum.'[31] One pendulum was suspended from a spring, the other on some form of knife-edge, presumably for experimental reasons. Greenwich Mean Time was in use.

There were plenty of teething troubles, as we might expect from clocks of such revolutionary design. The movements were mounted in the open behind the wainscot so they got dirty very quickly (Tompion would not let Flamsteed have the key). There was no form of temperature compensation on the long pendulums. Dust and swarf, strong winds shaking the house, moist weather, cold weather, hands sticking, wanting oil – these were the various reasons given by Flamsteed for bad going or stopping. Tompion had other ideas and blamed Towneley's escapement, replacing it with one of his own design.

Despite these horological troubles Flamsteed was able to proceed with his experiments into whether or not the Earth's rotation was 'isochronical' – whether or not it rotated at a constant speed. However, there were complications, one of which was caused by the fact that the Sun itself is not the most perfect of timekeepers. Because of the tilt of the Earth's axis and the fact that its orbit is not exactly circular (which means that the Sun is nearer to the Earth in the northern winter than in the summer), solar days vary in length

Edmond Halley was the official editor of what Flamsteed regarded as the unauthorized and inaccurate first edition of his *Historia Coelestis*, 1712. Halley succeeded Flamsteed as Astronomer Royal in 1720 and lived to give early support to John Harrison's work on finding longitude using an accurate timekeeper. From a portrait by Sir Godfrey Kneller.

through the year. The difference from the mean – the difference between clock time and sundial time on any day – is today called the Equation of Time but was known by Flamsteed as the Equation of Natural Days. He had recently published a new and more accurate table of the Equation, the existence of which was known to the Greeks but had still, in the seventeenth century, to be accurately quantified.[32] Time by the stars – sidereal time – does not suffer from these complications. Therefore, Flamsteed erected a telescope on the balcony so fixed that the transit of the bright star Sirius could be observed day and night throughout the year (unless there were clouds). The interval between two successive transits gave Flamsteed the precise length of one sidereal day, which, with a little arithmetic, he could use to check the clocks which were keeping Mean Solar Time.

As can be seen in the following extract from a letter from Flamsteed to Towneley dated 12 July 1677, the former's troubles were by no means all horological ones:

> I told you in one of mine formerly that our clocks kept so good a correspondence with the heavens that I doubted not but they would prove the Revolutions of the earth to be Isochronicall which if granted it will follow that the aequation of days which I demonstrated in the diatriba[33] is altogether agreeable to the heavens. I can now make it out by 3 moneths continued observations tho to prove it fully it is requisite the Clocks be permitted to goe a whole yeare without any alterations. . .
>
> If God send mee life to see them goe a yeare I doubt not but I shall be able to make out something more by them then is expected, if the office [of Ordnance] sterve mee not out, for my allowance you know is but small and now they are 3 Quarters in my debt, I feare I must come downe to the country to seeke some poore vicarage, then farewell to our experiment. . .[33]

Flamsteed seems to have been justified in his assertions. Reporting to his patron, Jonas Moore, on 7 March 1678, he says: 'My theory of the Equation of dayes I looked upon but as a dreame at first because one part on which it was founded viz the Isochroneity of the earths revolutions was onely supposed, not demonstrated, by me, but your clocks have proved that rational conjecture a very truth. . .'[34] So one of the ingredients in the solution of the problem of longitude at sea had been provided – by means of Greenwich time: it had been proved that the Earth rotates on its axis at a constant speed. As we shall see in Chapter Seven, it has now been proved that it does not – quite; but for practical navigational purposes, Flamsteed's results were adequate and stood for 250 years.

Presented to Flamsteed personally by Jonas Moore and removed from the observatory by Mrs Flamsteed in 1720 after her husband's death, the two Great Clocks still survive – one now being in the British Museum, the other back in the Great Room – today's Octagon Room – at Greenwich, where there are also replicas of the originals behind the wainscot. Their full story and that of Flamsteed's other clocks is told elsewhere.[35]

Publication of results

Though not directly connected with time, the rest of Flamsteed's doings are very much part of the longitude story, so deserve to be told, if only in brief. After the initial foundation, Charles II seems to have taken no further interest in his observator's affairs, which were directed by Jonas Moore until the latter's death in 1679. The Board of Ordnance undertook to keep the house in repair, pay Flamsteed a salary of £100 a year and another £26 for an assistant. However, the Astronomer Royal had to find his own instruments and pay for any skilled assistance he might need. To make ends meet, he was forced to take in private pupils.

Flamsteed was the first astronomer to use telescopic sights systematically for all his measurements. His earliest major instrument was an equatorially mounted sextant of 7 feet (2.13 metres) radius, the frame made by the Tower smiths and paid for by Jonas Moore who had also provided the two clocks. Between 1676 and 1690, 20,000 observations were taken with the sextant. In 1684 Flamsteed was presented to the living of Burstow in Surrey and in 1688 his father died. His improved circumstances enabled him to commission a new fundamental instrument, a 140° mural arc of the same radius as the sextant. This instrument, which cost him £120 and fourteen months' work, was mounted on a meridian wall of brick, enabling him to measure directly the co-ordinates of the heavenly bodies as they crossed the meridian. Before he died in 1719 more than 28,000 observations had been taken with the mural arc.

One consequence of the Government's parsimony was that Flamsteed regarded his astronomical results as his own property, to be published as and when he chose. This was the principal cause of the quarrel between Flamsteed on the one hand and Newton and Halley on the other, which was to bedevil the last twenty-five years of Flamsteed's life. In 1704 Prince George of Denmark, Consort of Queen Anne, agreed to stand the cost of publication of Flamsteed's observations, to be supervised by 'referees', including Sir Isaac Newton, who was by then President of the Royal Society. Publication proceeded very slowly and stopped altogether on the Prince's death in 1708. Relations between Flamsteed and the late Prince's referees deteriorated.

Since the death of Sir Jonas in 1679, Flamsteed had in effect been responsible directly to the Monarch, and none of them – Charles II, James II, William and Mary, Anne – seems to have taken any interest in their Royal Astronomer (the term Astronomer Royal was very seldom used at that time). In practice, Flamsteed was pretty well his own master.

On 12 December 1710, however, Queen Anne was persuaded to appoint a Board of Visitors to direct the affairs of the observatory, this Board to consist of the President – none other than Flamsteed's *bête noire* Newton – and other members of the Council of the Royal Society. Flamsteed was very angry and managed to stave off any visitation for the time being.

In 1711 Queen Anne gave orders that publication of his observations should proceed. The next year *Historiae Coelestis . . . observante Johanne Flamsteedio* was published in a single volume, edited by Edmond Halley, by now Flamsteed's sworn enemy. When Flamsteed saw the new book he was furious. All but 97 sheets had been printed without his having seen them. Material purporting to be Flamsteed's own was in fact only an abridgement of his results. He was particularly angry about the catalogue of star positions, which was full of errors, and about Halley's preface which was personally offensive.

On Saturday 1 August 1713 Flamsteed had to suffer his first (and only) 'visitation' from the Queen's representatives, who included Newton and Halley. Having given them a glass of wine he excused himself, pleading lameness (he was 67), saying they could go anywhere in the observatory, except for his library.[36]

Just before her death on 1 August 1714, Queen Anne gave her Royal Assent to the Longitude Act. In it Flamsteed was *ex officio* appointed one of the Commissioners for Longitude. However, though one or two hare-brained schemes were submitted to him for a professional opinion, there was no formal meeting of the Board of Longitude – as the Commissioners became known – during his term of office.

The Queen's death, however, benefited Flamsteed, at least in his quarrel with Newton. With the change of government from Tory to Whig, Flamsteed once again had friends at Court. On 28 March 1716, by King George I's order, 300 copies out of the 400 printed of Halley's edition of *Historiae Coelestis* were placed in Flamsteed's hands. Thriftily removing the 97 sheets which had his approval, he had the rest burned 'as a sacrifice to truth' in April 1716.[37] But he saved a few copies for those of his friends that were 'hearty lovers of truth, that you may keep them by you as evidences of the malice of godless persons, and of the candor and sincerity of the friend that writes to you, and conveys them into your hands: for l will not say

I make you a present of that which is so odious of itself, and will be detested by every ingenuous man.'[38]

These were strong words, but were they really justified? It is difficult to decide who was in the right in these rather tedious quarrels. Though there was some reason for Flamsteed's attitude towards the English scientific establishment, he had a duty, as a civil servant, to co-operate with others – or would by today's standards. Many of the wrongs he imputed were imagined ones and his prickliness, due at least in part to his chronic ill health, must have made any dealings with him frustrating in the extreme.

After the publication of what came to be known as Halley's 'pirate' edition of 1712, Flamsteed resolved to print his observations at his own expense. Before he died on 31 December 1719, all of Volume l – observations 1669-88, including the 97 sheets he had saved from the fire of 1716 – and most of Volume II – observa-

tions 1689-1719 – had been printed. The completion of Volume II and the whole of Volume III – containing a Latin preface[39] and his British Catalogue of star positions (thereby fulfilling King Charles's 1675 directive 'to the rectifying . . . the places of the fixed stars') – was undertaken by his two former assistants Abraham Sharp and Joseph Crosthwait who, incidentally, never got paid for their work.

The three volumes of Flamsteed's *Historia Coelestis Britannica* were eventually published in 1725, followed in 1729 by his *Atlas Coelestis*, a star atlas which, in effect, put his British Catalogue in graphic form. These were indeed monuments to fifty years' labour by a great astronomer.

Flamsteed's star catalogue, from his *Historia Coelestis Britannica*, Vol.III, published posthumously in 1725.

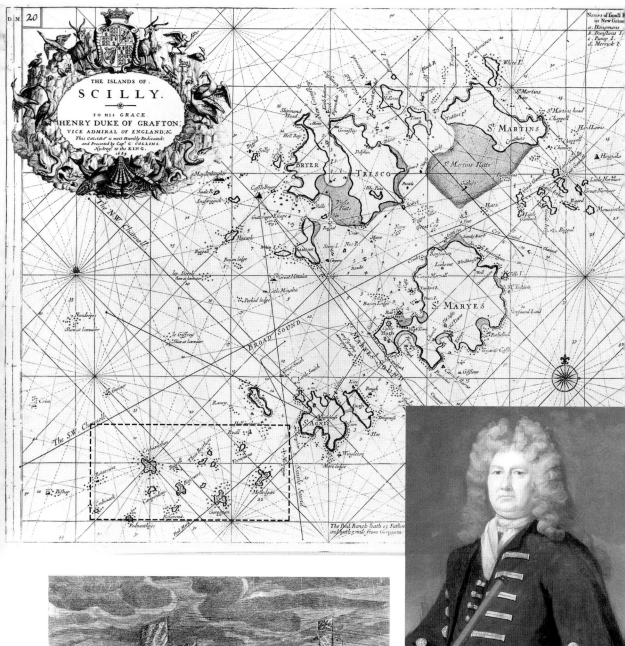

THE ISLANDS OF .
SCILLY.
TO HIS GRACE.
HENRY DUKE OF GRAFTON,
VICE ADMIRAL OF ENGLAND;&c.
This CHART is most Humbly Dedicated;
and Presented by Capt. C. COLLINS;
Hydrog. to the KING.

Top. Captain Greenvile Collins's
chart of the Isles of Scilly, 1698.
Shovell's fleet was lost in the area
indicated.
Above. Sir Clowdisley Shovell
(1650-1707), detail from a portrait
by Michael Dahl.
Left. The loss of Shovell's fleet, 22
October 1707. Anonymous
engraving.

THREE

Greenwich time for navigators
1700 - 1840

O N 29 September 1707 Admiral Sir Clowdisley Shovell with twenty-one ships of the Royal Navy sailed from Gibraltar for England. Britain was at war with France and, though Gibraltar had fallen into British hands in 1702, it was considered unwise for the larger ships of the fleet to remain in the Mediterranean during the winter. Weather on passage was not good. There were westerly gales from the 5th to the 10th of October. The 12th and 13th were squally, followed by two days of light winds. On the 16th and 17th there were easterly gales and on the 19th north-westerly gales. On the 21st the sky cleared and several ships took observations for latitude, while sounding gave a depth of between 90 and 140 fathoms, showing that the fleet was on the edge of the Continental Shelf. Three ships having been detached to Falmouth for convoy duty on the 22nd, and no sun being visible all day, the remaining 18 ships hove to in the afternoon to obtain soundings. Then, satisfied that they were in the mouth of the English Channel and clear of all danger, the ships ran to the eastward before a favourable gale.

About 7.30 p.m. the same day the *Association*, *Eagle* and *Romney* struck the Gilstone Ledges in the Scilly Isles. Of the 1,200 men in the three ships, only one man was saved - the quartermaster of the *Romney*. Sir Clowdisley's body was found floating the following day and now lies in one of the largest and ugliest tombs in Westminster Abbey. The *St George* and *Firebrand* also struck the rocks but got off. The latter, however, was badly holed and sank later, leaving only twenty-three survivors.[1]

This disaster – four ships were lost, with nearly two thousand men – was a profound shock to the British public. There had recently been other disasters: in 1691 several men-of-war off Plymouth were wrecked through mistaking the Deadman for Berry Head; in 1694 Admiral Wheeler's squadron, leaving the Mediterranean, ran aground on Gibraltar when they thought they had passed the Strait. And there were soon to be more. In 1711 several transports were lost near the St Lawrence river, having erred 15 leagues in their reckoning during twenty-four hours; and in

William Whiston, who, with Humphrey Ditton, petitioned the House of Commons in April 1714, resulting in the passing of the Longitude Act. Detail from an oil painting by an unidentified artist.

1722 Lord Belhaven was lost on the Lizard the same day on which he had sailed from Plymouth.[2]

A recent analysis of the Clowdisley Shovell disaster concluded that, while uncertain currents, fog and bad compasses might have contributed to the disaster, much more must be attributed to the lack of accurate knowledge of the geographical positions of the headlands concerned, to bad charts, to bad navigation textbooks, and to the generally low standard of accuracy of navigational practice at that time.[3] Though this disaster was not actually caused by the lack of a method for finding longitude at sea, nevertheless its very magnitude made such an impression on the British public that they became more than ever receptive to any suggestion that might make navigation safer – and in the 1710s 'finding the longitude' seemed to hold the key to this. King Charles's foundation of the Royal Observatory was obviously not enough (probably less than one in twenty of the population knew about that anyway). Something more must be done by the Government of his niece, Queen Anne, then in the throes of a ministerial crisis.

On 14 July 1713 there appeared a letter in the journal *The Guardian*, introduced by the editor Joseph Addison with these words: '. . . It [the letter] is on no less a subject than that of discovering the longitude, and deserves a much higher name than that of a Project, if our language afforded any such term. But all l can say on this subject will be superfluous when the reader sees the names of those persons by whom this letter is subscribed, and who have done me the honour to send it to me . . .'[4] The writers of the letter which followed were William Whiston (1667-1752) and Humphrey Ditton (1675-1714), Mathematical Master at Christ's Hospital. In 1703, after several years as a country vicar, Whiston had succeeded Sir Isaac Newton as Lucasian Professor of Mathematics at Cambridge, but his theological views led to his expulsion from the university in 1710. Whiston then set up as a lecturer on scientific and religious subjects, the first being a series on astronomy given at Button's Coffee House with the encouragement of Addison and Steele.[5] The letter started:

> It is well known, Sir, to yourself and to the learned, and trading, and sailing worlds that the great defect of the art of navigation is that a ship at sea has no certain method in either her eastern or western voyages, or even in her less distant sailing from the coasts to know her Longitude, or how much she has gone eastward or westward, as it can easily be known in any clear day or night, how much she is gone northward or southward.[6]

Having surveyed possible methods of finding longitude, Whiston and Ditton went on to say that they had a method of their own, one which would provide not only longitude but latitude as well, which they would disclose if they were offered a reward, subject to a satisfactory report by Sir Isaac Newton and other persons.

The method eventually disclosed, in a book published in 1714,[7] turned out to be a proposal that vessels should be moored in known positions at intervals along the trade routes, each fitted with a mortar which would, every midnight by Peak of Tenerife time (Whiston and Ditton's proposed prime meridian), fire vertically a projectile, some sort of tracer or rocket visible from afar, which would culminate – or perhaps burst – at 6,440 feet precisely. Ships would look out for these projectiles at midnight. A compass bearing would give the direction from the recorded position of the lightship. The distance of the viewer from the lightship could be obtained either by noting the difference in time between seeing the flash of the gun's discharge and hearing the report, or by measuring the elevation of the highest point of the shell. This proposal was still more fantastic because the light-vessels in deep water were to be moored by lowering some form of sea-anchor through the upper layers of water to the allegedly immovable layers beneath.

Ode, for Musick. On the Longitude

Recitativo
The Longitude mist on
By wicked *Will Whiston*
And not better hit on
By good Master *Ditton*.

Ritornello
So *Ditton* and *Whiston*,
May both be bep-st on;
And *Whiston* and *Ditton*
May both be besh-t on.

Sing *Ditton*,
Besh-t on;
And *Whiston*
Bep-st on.

Sing *Ditton* and *Whiston*,
And *Whiston* and *Ditton*,
Besh-t and Bep-st on,
Bep-st and Besh-t on.

Da Capo[9]

Bizarre though this scheme may now appear, it seems in general to have been taken seriously by some of Whiston and Ditton's contemporaries. Not so the Scriblerus Club, however, that club devoted to satirizing false learning. The group of wits comprising the club – Dr. Arbuthnot, Alexander Pope, John Gay, Jonathan Swift, Thomas Parnell and Queen Anne's Lord Treasurer, the Earl of Oxford – were convulsed by the publication of Whiston and Ditton's long-awaited book. Arbuthnot wrote to Swift on 17 July 1714, a few days after the book appeared: 'Whiston has at last published his project on the longitude; the most ridiculous thing that was ever thought on. But a pox on him! He has spoiled one of my papers of Scriblerus, which was a proposal for the longitude not very unlike his . . . that all the Princes of Europe join to build two prodigious poles, upon high mountains, with a vast lighthouse to serve for a pole star.'[8] Some time afterwards, a set of unsavoury verses began to circulate (*see left*).

The authorship of these verses has long been the subject of debate in bibliographical circles. Published in 1727 in the 'Last' Volume of Pope's and Swift's *Miscellanies*, they were ascribed first to Pope, then to Swift. Joseph Spence attributed the verses to Gay in 1820. However, the question seems to be settled by the following extract from a letter dated 23 December 1714 from Sir Richard Cox, former Lord Chancellor of Ireland, in Dublin, to Edward Southwell, Secretary of State in Whitehall, London: 'Archdeacon Parnell has made the following dirty lines, which are valued, because they Ridicule the Confident Arrian Whiston [Whiston had a reputation for holding Arian beliefs].'[10] Then follow, under the title 'A Round O on the Longitude by Whiston & Ditton', a text shorter than, and varying in details from, that quoted above.

Whiston and Ditton were nevertheless encouraged by the feelings of the public at large on the subject of longitude and presented a petition to Parliament at the end of April 1714, begging for a reward for 'discovering the longitude' – and submitting their own scheme. The House of Commons Journal for 25 May reprinted another petition a month later:

A Petition of several Captains of her Majesty's Ships, Merchants of *London*, and Commanders of Merchantmen, in behalf of themselves, and all others concerned in the Navigation of *Great Britain*, was presented to the House, and read; setting forth, That the Discovery of the Longitude is of such Consequence to *Great Britain*, for Safety of the Navy, and Merchant Ships, as well as Improvement of Trade, that, for want thereof, many Ships have been retarded in their Voyages, and many lost; but if due Encouragement were proposed by the Publick, for such as shall discover the same, some Persons would offer themselves to prove the same, before the most proper Judges, in order to their entire Satisfaction, for the Safety of Mens Lives, her Majesty's Navy, the Increase of Trade, and the Shipping of these Islands, and the lasting Honour of the *British* Nation: And praying their Petition may be taken into Consideration.[11]

This petition was sent to a Committee of the House who in turn co-opted technical experts, including Sir Isaac Newton, President of the Royal Society, and Edmond Halley who later succeeded Flamsteed as Astronomer Royal. Sir Isaac gave a summary of the current position, reported thus to the House on 11 June:

> Sir *Isaac Newton*, attending the Committee, said, That, for determining the Longitude at Sea, there have been several Projects, true in the Theory, but difficult to execute:
>
> One is, by a Watch to keep Time exactly: But, by reason of the Motion of a Ship, the Variation of Heat and Cold, Wet and Dry, and the difference of Gravity in different Latitudes, such a Watch hath not yet been made:
>
> Another is, by the Eclipses of *Jupiter's Satellites*: But, by reason of the Length of Telescopes requisite to observe them, and the Motion of a Ship at Sea, those Eclipses cannot yet be there observed:
>
> A Third is, by the Place of the Moon: But her Theory is not yet exact enough for this Purpose: It is exact enough to determine her Longitude within Two or Three Degrees, but not within a Degree:
>
> A Fourth is *Mr Ditton's* Project: And this is rather for keeping an Account of the Longitude at Sea, than for finding it, if at any time it should be lost, as it may easily be in cloudy Weather: How far this is practicable, and with what Charge, they that are skilled in Sea-affairs are best able to judge: In sailing by this Method, whenever they are to pass over very deep Seas, they must sail due East or West, without varying their Latitude; and if their Way over such a Sea doth not lie due East, or West, they must first sail into the Latitude of the next Place to which they are going beyond it; and then keep due East, or West, till they come at that Place:
>
> In the Three first Ways there must be a Watch regulated by a Spring, and rectified every visible Sun-rise and Sun-set, to tell the Hour of the Day, or Night: In the Fourth Way, such a Watch is not necessary: In the First Way, there must be Two Watches; this, and the other mentioned above:'[12]

The House accepted the report unanimously and directed that a Bill should be prepared by several Members, among whom were Joseph Addison and General James Stanhope, later the first Earl. *A Bill for Providing a Publick Reward for such Person or Persons as shall Discover the Longitude at Sea* was presented to the Commons on 16 June. It offered rewards of unprecedented magnitude to '. . . the First Author or Authors, Discoverer or Discoverers of any such Method . . . To a Reward, or Sum of Ten Thousand Pounds, if it Determines the said Longitude to One Degree of a great Circle, or Sixty Geographical Miles;* to Fifteen Thousand Pounds, if it Determines the same to Two Thirds of that Distance; and to Twenty Thousand Pounds, if it Determines the same to One half of

---

*A geographical mile is the distance subtended by an arc-minute on the equator, or 6,087 feet, very close to the nautical mile whose length, however, varies with latitude because the Earth is not a perfect sphere. The English statute mile is 5,280 feet.

the same Distance . . .'[13] A half of the reward was to be paid as soon as the Commissioners were satisfied that 'any such Method extends to the Security of Ships within Eighty Geographical Miles of the shores, which are the Places of greatest Danger', the other half to be paid 'when a Ship by the Appointment of the said Commissioners, or the major part of them, shall thereby actually sail over the Ocean, from Great Britain to any such Port in the West-Indies, as those Commissioners, or the major part of them, shall Choose or Nominate for the Experiment, without losing their Longitude beyond the limits before mentioned'.

No two economic historians agree about the value in today's money of the £20,000 in 1714, but almost all put it above a million pounds, while some put it as high as £3.5m.[14] A prize indeed! But the Bill went on to stipulate that, before the reward could be paid, '... such Method for the Discovery of the said Longitude shall have been tried and found Practicable and Useful at Sea . . .' The definition of 'practicable and useful' was to cause much acrimony in future years.

There were two further provisions of significance in the Bill: first, that the Commissioners were empowered to advance sums of up to £2,000 for promising schemes, 'to make experiment thereof'; secondly, if a proposal on trial did not quite match up to the specifications for the main rewards, but was nevertheless found to be 'of considerable Use to the Publick', then it could qualify for some lesser reward, to be valued at the Commissioners' discretion. The provisions of the Bill applied to all who satisfied the conditions, regardless of nationality.

The Bill was read for the first time on 17 June 1714, passed by the Commons on 3 July, and passed by the House of Lords on 8 July. Queen Anne gave her Royal Assent on 20 July, only twelve days before her death.

The Commissioners appointed by the Act, who came to be known as the Board of Longitude, were the Lord High Admiral of Great Britain, or First Commissioner of the Admiralty; the Speaker of the House of Commons; the First Commissioner of the Navy; the First Commissioner of Trade; the Admirals of the Red, White and Blue Squadrons; the Master of Trinity House; the President of the Royal Society; the Astronomer Royal; the Savilian, Lucasian and Plumian Professors of Mathematics at Oxford and Cambridge Universities; and ten named Members of Parliament.

The immediate effect of the passing of the Longitude Act was to stimulate the publication of many pamphlets by those who thought they had solutions to the longitude problem, several of which are quoted by Gould.'[15] Two proposals have survived among the papers of John Flamsteed who, as we have seen, was one of the

Commissioners appointed by the Act: the first was by William Hobbs, 'c/o Mr Jam. Hubert's watchmaker, in Finch Lane near the Royal Exchange', for his 'Movement with a Decimal Horloge';[16] another by Digby Bull of Three Logs Court in White Cross Street, London, for his 'shipwatch'.[17] Both asked for trials to be carried out but neither received any recognition.

Even the eminent Sir Christopher Wren, then aged 82, put in a claim. The following paper was found in the nineteenth century among the Newton manuscripts at the Royal Society:

> Sir Christopher Wrenn's Cypher, describing three Instruments proper for discovering the Longitude at Sea, delivered to the Society Novemb. 30, 1714, by Mr. Wren [the son]:
>
> OZVCVAYINIXDNCVOCWEDCNMALNABECIRTEWNGRAMHHCCAW.
> ZEIYEINOIEBIVTXESCIOCPSDEDMNANHSEFPRPIWHDRAEHHXCIF.
> EZKAVEBIMOXRFCSLCEEDHWMGNNIVEOMREWWERRCSHEPCIP.
>
> Vera Copia – Edm: Halley[18]

This was deciphered by Francis Williams of Chigwell, Essex, and reported at the British Association meeting of 1859:

(a) Reverse each line end for end.
(b) Strike out every 3rd letter in each line: these give names and dates.
(c) The residue is the text, thus:

3rd letters	*Residue*
CHR WREN MDCCXIV Z	WACH MAGNETI BALANCE
	WOVND IN YACVO = Watch-Magnetic balance wound in vacuo
CHR WREN MDCCXIIII	FIX HEAD HIPPFS HANDES
	POISE TVBE ON EYEZ = Fix head, hips, hands - poise tube on eyes
CHR WREN MDCCXIV Z	PIPE SCREWE MOVING
	WHEELS FROM BEAKE = Pipe screw moving wheels from beak[19]

A detailed explanation of this cryptogram has never been found, but the first phrase must refer to what would be called today a marine chronometer with a magnetic balance, the whole being kept in a vacuum, a measure advocated by Wren (erroneously, as it turned out) many years before to combat the effects of temperature changes on timekeeping. The second phrase seems to refer to Wren's ideas for observing Jupiter's satellites at sea, the main difficulty of which arises from the fact that the motion of the ship makes it impossible (as it has since been proved) to hold the telescope steady enough for accurate timing of eclipses. The third must refer to some kind of log for measuring a ship's speed through the water, not unlike that proposed by Reusner to Louis XIV in 1668.[20]

The cryptogram was a device commonly used in the best scientific circles of the time to establish priority of invention or discovery without actually disclosing anything that might be seized upon by a zealous colleague. It was used by Galileo, Huygens and Hooke, the last of whom published a Latin cryptogram giving (when translated) the principle of the balance spring in a watch: 'As the tension

Hogarth's longitude lunatic, seen here scribbling on the wall. Detail from Hogarth's *The Rake's Progress*, plate 8, 1st state, 1735.

is, so is the force' (that is, the force exerted by the spring is proportional to the amount of tension).

None of these early proposals came to anything. If the Board of Longitude met to consider any of them, no record has survived, the earliest meeting for which minutes are available being 30 June 1737, twenty-three years after its establishment. Meanwhile, exactly as had happened a little over a hundred years before in Spain, the phrase 'finding the longitude', coupled once again with squaring the circle (the Board received many suggestions for doing this, all impracticable and none really concerned with longitude), passed into the English language as expressing something which, if not downright impossible, was extremely difficult to achieve. It was used as a catch-phrase in newspapers and broadsheets. Gulliver (published 1726), describing what he would do if he became immortal like the Struldbrugs, says, 'I shall then see the discovery of the longitude, the perpetual motion, the universal medicine, and many other great inventions brought to the utmost perfection.'[21] In 1735, in the final, madhouse, scene in his series of paintings *The Rake's Progress*, Hogarth includes a man scribbling on the wall calculating the longitude, together with the religious maniac, the mad tailor, the mad astronomer, the mad musician, the man who fancies himself Pope, and other lunatics. And on the wall is a sketch of a ship with a mortar discharging a bomb vertically, a reference to the Whiston-Ditton proposal of 1714.[22] Even as late as 1773 Goldsmith makes the ingenuous Marlow say: 'Zounds, man! we could as soon find the longitude!' in reply to Tony Lumpkin's complicated (and fictitious) directions for finding Hardcastle's house.[23]

French prizes

In France the State did not follow the British example. However, on 12 March 1714 (two months before the longitude debates in the British House of Commons) Rouillé de Meslay, French parliamentary counsel, drew up a will in which he bequeathed 125,000 livres (at about 20 livres to the pound sterling) to found two prizes, to be awarded annually by the French Académie des Sciences for philosophical dissertations on two specific subjects of their choice. The larger prize was to be for a dissertation on the chosen subject concerning the make-up and motions of the solar system and of the principles of light and motion; the smaller was to go to him 'who best achieved the shortest and easiest method and rule for taking the heights and degrees of longitude at sea exactly, and [who made] useful discoveries for navigation and great voyages.'[24] Rouillé died in 1715 and the Académie accepted the bequest in March 1716. Despite attempts by his son to get the will annulled, the legacy was confirmed by the High Court in 1718. In the meantime another prize had been offered in 1716 by Philippe, Duke of Orléans and Regent of France, to be awarded by the Académie (the amount unspecified) to the inventor of the secret of finding the longitude. For reasons that are not known, the Académie never awarded such a prize.[25]

The first Rouillé prize was offered in 1720 for the solution to the problem of finding the best way to ensure the even going of a clock at sea, whether by the design of the movement itself, or by the suspension. The prize of 500 livres was won by a Dutch horologist called Massy for his proposals, on paper only, for a watch to be kept in a box maintained at an even temperature by a lamp.[26] In fact Rouillé's capital sum proved insufficient to allow the prizes to be awarded annually so, for the next seventy years, a single prize of 2,000 livres was offered every two years or more, as funds permitted. The second prize in 1725 today seems somewhat bizarre: 2,000 livres to Daniel Bernoulli, the Swiss scientist, for his answer to the question of how best to keep even movement at sea of clepsydras (water-clocks) and sand-glasses. Later prize questions concerned compasses, the motion of the Moon, measurement of altitude, the best height of masts; while others concerned marine timekeepers.[27]

The invention of the sextant

Before returning to the longitude problems as such we must take note of the development of angle-measuring instruments, because, as explained in Appendix 1, the lunar-distance method of finding longitude at sea required the navigator to measure very precisely both the angle between the Moon and the Sun or star, and the altitude of both bodies above the horizon. Werner in 1514 had proposed the use of a cross-staff for this purpose. However, astronomers soon realized that such an instrument would never have the necessary accuracy. Assuming a lunar distance could be measured to an accuracy of half a degree (which with a cross-staff it probably could not), then this would be equivalent to an uncertainty of one hour in finding Greenwich Time, or of 15° in longitude. But what was needed was to find longitude to 1° or better, which demanded a lunar-distance-measurement accuracy of at least two minutes of arc.

The first new idea of any promise was reported to the Royal Society in London by Robert Hooke who, early in 1666, announced that he was preparing 'a perspective for observing positions and distances of fixed stars from the Moon by reflection', explained by him with drawings later the same year. In 1691 Halley produced designs for an instrument which was very much the same, so much so that

Halley, under pressure from Hooke, withdrew and admitted the latter's priority of invention.[28] There is no record of either of these instruments having been tried at sea.

Then came two other developments, one abortive and one fruitful. In 1729 the Académie offered the third Prix Rouillé for the best methods of observing the altitude of the Sun and stars at sea, by instruments known or to be known. An award of 2,000 livres was made to Pierre Bouguer, professor of hydrography at Croissic, for his design of a back-staff in the form of a quadrant where the observer could see both Sun and horizon simultaneously.[29] However, as it worked on a shadow principle, it could not be used for lunar distances. The second event was no less than the simultaneous invention in 1731 of a double-reflection quadrant, the ancestor of today's sextant, by two people quite independently on both sides of the Atlantic. John Hadley (1682-1744), vice-president of the Royal Society, who had already made many improvements to the reflecting telescope, produced designs for two instruments on a principle which was thought to be entirely new, that of double reflection, the ray of light from the heavenly body reaching the eye after being reflected twice, once by a mirror attached to a moving index-arm, then by a mirror fixed in relation to the sighting vane. This allowed both bodies (for lunar distances) or the body and the horizon (for altitudes) to be seen by the observer simultaneously, making observations in a moving ship practicable. Hadley described these instruments in a communication to the Society in May 1731 and they were tested at sea in the *Chatham* yacht in the Thames estuary from 31 August to 1 September 1732, the errors of altitudes and of the distances between two stars obtained by the second of the two instruments he had described proving to be less than 2 arc-minutes, well within the requirement for lunar distances.[30]

Then, in 1732, the Royal Society received a letter through Edmond Halley from James Logan, Chief Justice of Pennsylvania, enclosing affidavits purporting to prove that Thomas Godfrey (1704-49), a glazier and self-taught astronomer from Philadelphia, had designed a double-reflection instrument capable of observing lunar distances substantially the same as the first of the two designed by Hadley. When Benjamin Franklin started in business in Philadelphia with Hugh Meredith he sub-let part of his house to Godfrey, who became one of the founder-members of the

Taking a 'lunar', sketched by J.L. Kirby, second officer of the Blackwall frigate *Owen Glendower*, on passage from England to Bombay, 1846-47.

Leathern-apron Club, or Junto, a debating club founded by Franklin, devoted to morals, politics, and natural philosophy. Franklin, though he called Godfrey 'a great mathematician', thought him something of a dull stick, describing him as not a pleasing companion.[31] Godfrey calculated ephemerides for an almanac which Franklin printed and distributed, and he also assisted Logan in making astronomical observations at his home.[32] Godfrey's instrument had been tried in the *Trueman* sloop on voyages to Jamaica and Newfoundland in 1730-1, proving highly successful, whereupon Godfrey had shown the design to Logan, asking him to transmit it to the proper authorities in the hope of receiving an award under the Longitude Act. Unfortunately Logan took no action until he saw Hadley's description in the *Philosophical Transactions*. The Royal Society investigated the claims of both parties in January 1733 and came to the conclusion that this was indeed a near-simultaneous invention by Godfrey and Hadley.

But this was not the end of the story. In 1742 there was found among Edmond Halley's papers after his death a drawing and description of a double-reflection quadrant almost identical in its general arrangement with those of Godfrey and Hadley, designed by no less a person than Sir Isaac Newton. Apparently Newton had shown the design to Halley in 1700 but the latter felt there was no merit in it. Though the principle was the same as that used in a nautical sextant today, the details of Newton's actual design make it unlikely that it would have been feasible at sea. Nevertheless, had his invention been made public earlier, a practicable nautical angle-measuring instrument might have been available some years sooner than it actually was.

Whoever may have been the true inventor, the instrument which was to prove the basis for all subsequent developments for angle-measurement at sea almost up to the present day came to be known as Hadley's reflecting quadrant – or just plain 'Hadley' – and, over the next twenty years or so, appeared on the commercial market all over the world, superseding all previous instruments of that kind. Three subsequent developments are worth mentioning. First, in 1752 Tobias Mayer, an astronomer from

Tobias Mayer (1723-62), a surveyor and professor of astronomy at the University of Göttingen, in Germany, was the first to develop lunar and solar tables of sufficient accuracy for finding longitude by observational means.

TOBIAS MAYER

Göttingen, whom we will be meeting again, introduced a repeating circle, an instrument able to measure angles of any size and having other theoretical advantages as far as precision was concerned: circular instruments, however, never became popular among British seamen, being both cumbersome and expensive. Secondly, about 1757 the English instrument-maker John Bird and navigator Captain John Campbell developed the sextant proper, with an arc of 60° measuring angles up to 120° (instead of Hadley's arc of 45° measuring to 90°), with a frame of brass more rigid than previous wooden frames, and with telescopic sights. Finally, Jesse Ramsden's 'dividing engine' of 1775 allowed sextants and circles to be made smaller and lighter with no loss of accuracy.

So the navigator now had an instrument with which he could measure lunar distances with adequate precision at sea. Another part of the lunar-distance problem had been solved, another ingredient found.

Sir Isaac Newton (1642-1727), President of the Royal Society, who had designed a double-reflection quadrant by 1700. An oil painting of c. 1860, after Sir Godfrey Kneller.

The Nautical Almanac

The problem of how to find longitude at sea astronomically – which gave rise to the need by navigators to find or keep Greenwich Time (or maybe Paris Time, or Cadiz Time) – was being solved piece by piece. Flamsteed had cleared the air by proving that, for practical navigational purposes, the Earth rotates at a constant speed. The Jupiter's satellite method seemed unlikely ever to be practicable at sea. The chronometer method, which was ultimately to prove the best, still awaited the development of a timekeeper which could keep precise time for months on end in any climate, regardless of the motion of the ship: but there were indications that such an invention would not be long in coming.

For the lunar-distance method, to provide data for which Charles II had founded Greenwich Observatory, Flamsteed had produced an adequate catalogue of star positions, while Hadley's reflecting quadrant promised to give the necessary accuracy for the actual measurement of lunar distances. What was still needed was an adequate theory of the complicated motion of the Moon so that her position against the background of the stars could be predicted several years in advance. And all these data had to be presented to the navigator in such a form that he could actually use them to find his longitude without too much labour.

This particular part of the longitude and time story – how the theory of the Moon's motion came to be established to a degree of accuracy useful to navigation – is highly technical and we will not attempt to describe here *what* was discovered, but will merely mention the leading characters concerned.

It was Newton's theory of gravitation, set out in his *Principia* of 1687, that first offered some hope of explaining the irregularities of the Moon's motion. However, as Anton Shepherd was to say later, '... but still, for want of a more continued and uninterrupted Series of Observations of the Moon, than those of Mr *Flamsteed*, the Difference of Sir *Isaac's* Theory from the Heavens would sometimes amount at least to five Minutes [of arc]'[33] – which was the equivalent of an error of 2½ in the longitude found. In the next fifty years or so, so important was it to 'find the longitude' that the world's best mathematicians turned their minds to producing the theory, and the world's best astronomers, particularly those at Greenwich, to producing the data needed to predict the motion of the Moon. Edmond Halley was appointed to succeed Flamsteed as Astronomer Royal at Greenwich in 1720. As we have seen, Flamsteed had concentrated upon the stars: Halley determined to concentrate upon the Moon.

Every 18 years 11.3 days, the so-called eclipse or Saros cycle of 223 lunations (intervals between new moons), the motions of the Moon relative to those of the Sun repeat themselves. Halley reasoned that the best way to predict the Moon's position today was to know by measurement her position 223 lunations ago. Therefore in 1722, at the age of 66, he set himself the task of observing the position of the Moon on every possible occasion when she was visible crossing the meridian (that is, except for New Moon and clouds) throughout the 18-year cycle, thus confidently setting himself a task that could not be completed until he was 84. He announced this in the *Philosophical Transactions* of 1731. As there was no instrument then accurate enough to measure lunar distances at sea, he first recommended that the navigator should make use of that special case of the lunar distance when the Moon actually passes in front of, or very close to, the star (the technical terms for which are occultation and appulse), requiring only a telescope to observe. As a postscript to his paper he mentioned that his fellow vice-president of the Royal Society, John Hadley, in that very year, 'has been pleased to communicate his most ingenious Invention of an Instrument for taking the Angles with great Certainty by Reflection . . . it is more than probable that the same may be applied to taking Angles at Sea with the desired Accuracy.'[34]

Taking a lunar distance.
From E. Dunkin, *The Midnight Sky*, 2nd edn.
(London [1879]), p.256.

TAKING A LUNAR DISTANCE.

Halley died in 1742 at the age of 86. Though he had no regular observational help, he nearly accomplished his aim of observing the Moon throughout the eighteen-year cycle. Alas, the accuracy of his observations – he was over 60 when he started, remember – left much to be desired and, when his lunar tables were eventually published, observations soon proved his predictions wrong. Meanwhile on the Continent the mathematicians and astronomers Lemonnier, Cassini de Thury, Euler, D'Alembert,and Clairaut were all engaged on the problem and, stimulated by prizes offered in 1750 and 1752 by the Academy of St Petersburg, various new theories of motions and lunar tables were published; but all carried errors of 3 to 5 minutes of arc in the place of the Moon, possibly because they were based on too few observations.[35]

It was a practical astronomer who finally came up with lunar tables of the accuracy required – Tobias Mayer of Göttingen, inventor of the repeating circle. Using Euler's equations, he produced tables of the Sun and Moon based on his own observations and those of James Bradley (1693-1762), who had succeeded Halley at Greenwich in 1742. In 1755 he sent a memorial to Admiral Lord Anson, First Lord of the Admiralty in London, enclosing his tables. They were laid before the Board of Longitude on 6 March 1756

[128] OCTOBER 1772.

Distances of ☽'s Center from ☉, and from Stars west of her.

Days	Stars Names	Noon.	3 Hours.	6 Hours.	9 Hours.
		D. M. S.	D. M. S.	D. M. S.	D. M. S.
1		62. 6.55	63.44.49	65.22.18	66.59.22
2		74.58.25	76.32.59	78. 7.10	79.40.56
3	The Sun.	87.24. 0	88.55.28	90.26.35	91.57.21
4		99.26. 2	100.54.47	102.23.14	103.51.23
5		111. 7.52	112.34.22	114. 0.37	115.26.38
3		33. 0.51	34.36.17	36.11.37	37.46.49
4	Antares.	45.40.30	47.14.40	48.48.39	50.22.24
5		58. 8. 6	59.40.36	61.12.55	62.45. 2
6		70.22.45			
6		15.30.17	17. 2.20	18.34.12	20. 5.52
7	β Capri-	27.41.48	29.12.32	30.43. 8	32.13.37
8	corni.	39.44.19	41.14. 8	42.43.52	44.13.31
9		51.40.38			
9		57.50.30	59. 8.51	60.27.28	61.46.19
10	α Aquilæ.	68.23.45	69.43.44	71. 3.50	72.24. 5
11		79. 6.58	80.27.49	81.48.43	83. 9.41
12		59.29.53	60.47.28	62. 5.21	63.23.34
13	Fomal-	69.58.45	71.18.29	72.38.25	73.58.33
14	haut.	80.41.47			
14		64.16.48	65.41.57	67. 7.19	68.32.59
15	α Pegaſi.	75.43.16	77. 9.54	78.36.43	80. 3.42
16		87.21.14			
16		43.43.39	45.11.59	46.40.39	48. 9.40
17	α Arietis.	55.39.44	57.10.44	58.42. 4	60.13.44
18		67.56.57	69.30.35	71. 4.33	72.38.50
19		47. 6.49	48.45. 7	50.23.45	52. 2.47
20	Aldeba-	60.23.26	62. 4.44	63.46.26	65.28.32
21	ran.	74. 5.12	75.49.46	77.34.46	79.20.10
22		46.12.29	47.56. 9	49.40. 7	51.24.49
23	Pollux.	60.16.20	62. 4. 6	63.52.16	65.40.51
24		74.49. 8	76.39.41	78.30.30	80.21.31
29		42.58.46	44.38. 3	46.16.55	47.55.19
30	The Sun.	56. 0.32	57.36.15	59.11.31	60.46.20
31		68.33.43	70. 5.53	71.37.39	73. 8.59
N.1		80.39.43			

Lunar-distance table from *The Nautical Almanac and Astronomical Ephemeris for the Year 1772*. This publication cut the navigator's arithmetical work for longitude computation from four hours or more to about half an hour.

with a recommendation from Bradley that they be tried at sea, 'but that, previous thereto, proper Instruments should be made to take the necessary Observations on Ship board, Hadley's Quadrant not being, in his opinion, altogether fit for that purpose'.[36] Because of the constraints imposed by the Seven Years War, which had broken out in 1756, the sea trials carried out by Captain Campbell in 1757-8 were not conclusive as far as Mayer's lunar tables were concerned. However, the trials proved most important for another aspect of the lunar-distance problem, in that it was Campbell's lunar-distance observations at sea, with a small-scale brass model of Mayer's own repeating circle, that led to the invention of the nautical sextant we know today.

It was not until 1761 that Mayer's tables were properly tested, by Nevil Maskelyne (1732-1811), the future Astronomer Royal, on his voyage to and from the island of St Helena where he was sent by the Royal Society to observe the transit of Venus. Using a Hadley quadrant (not a sextant) and Mayer's first tables, he made some successful lunar-distance observations, generally achieving an accuracy in longitude of better than 1°. Immediately on his return from St Helena, Maskelyne published his *British Mariners' Guide*, explaining in simple terms lunar-distance observations at sea.[37] In fact, Maskelyne was not the first person to use lunar distances successfully at sea. In 1753-4 the Abbé Nicolas-Louis de Lacaille had made such observations on his way home to France from the Cape of Good Hope by way of Mauritius and Réunion, though his lunar tables were less precise than those of Mayer: it was his method of working which Maskelyne used and recommended in his book. Also, in 1760 the Danish scholar Carsten Niebuhr made similar observations at Mayer's own request, while Alexandre-Guy Pingré followed Lacaille's example on his way to and from Rodriguez Island to observe the 1761 transit of Venus.[38]

Tobias Mayer died in 1762 but, shortly before, he had prepared a new and more accurate set of tables of the Sun and Moon, the

The Reverend Nevil Maskelyne was the leading promoter of the 'lunar distance' method of determining longitude. When Harrison submitted his H4 timekeeper for the longitude prize, Maskelyne did not believe that it met fully the winning criteria. His views became critical after his appointment as Astronomer Royal in 1765. Louis van der Puyl's portrait of 1785 shows him with the Observatory at Greenwich in the background.

manuscript of which was sent to England by his widow and considered by the Board of Longitude on 4 August 1763, when it was resolved that the new tables should be tried out by 'the Person who goes to Jamaica to make observations of Jupiter's Satellites'.[39] In the event, the person sent was Maskelyne himself – to Barbados, not Jamaica, principally in connection with the trials of Harrison's fourth timekeeper. However, he did also try out Mayer's new tables, using one of the newly invented sextants, with which he managed to fix the position of the Isle of Wight to within 16 arc-minutes of the true longitude.[40]

Bradley had died in 1762 and was succeeded as Astronomer Royal by Nathaniel Bliss who died only two years later. On 19

January 1765 the Board of Longitude heard from the Earl of Sandwich that King George III had agreed to the appointment of Maskelyne himself – returned only a few months before from Barbados – to succeed Bliss as Astronomer Royal, and he thereby became *ex officio* a Commissioner for Longitude. At their meeting of 9 February Maskelyne submitted a memorial reporting his successful lunar-distance observations for longitude during his voyages to and from St Helena and Barbados, and recommending that the Board publish a Nautical Ephemeris to make the whole matter of lunar observations easy for seamen: in this, he produced testimonials from four officers of the East India Company.[41]

Planned and executed by Maskelyne with characteristic energy – he was still only 33 – *The Nautical Almanac and Astronomical Ephemeris for the year 1767* was published late in 1766, and this remarkable publication has continued annually until the present day. In a letter to his brother Edmund in Calcutta, dated 15 May 1766, Nevil Maskelyne said:

> The board of longitude have engaged persons to compute a very complete nautical & astronomical ephemeris which will come out next Septr. for the year 1767: and be continued annually. There will be 12 pages in every month. All the lunar calculations for finding the longitude at sea by that method will be ready performed: & other useful & new tables added to facilitate the whole calculation; *so that the sailors will have little more to do than to observe carefully the moon's distance from the sun or a proper star,** which are also set down in the ephemeris, in order to find their longitudes . . .[42]

It is clear from this that the most practical features in the new almanac were the tables of lunar distances giving, for every three hours throughout the year, predicted angular distances from the Moon's centre to selected zodiacal stars, or to the Sun's centre. Such tables had been suggested by Lacaille to cut down the amount of arithmetical work needed by navigators. And so it proved: the time taken to work out a sight and obtain a longitude at sea was cut from over four hours to thirty minutes or so.[43]

In 1675 King Charles II had directed his royal astronomer 'to apply himself . . . to the rectifying the tables of the motions of the heavens, and the places of the fixed stars . . .' In 1725 Flamsteed's *Historia Coelestis Britannica* had published the places of the fixed stars. Now, ninety-one years after Charles's original warrant, the fifth royal astronomer Nevil Maskelyne published 'the tables of the motions of the heavens' in a form suitable for navigators '. . . so as to find out the so-much-desired longitude of places for the perfecting the art of navigation.'[44] So King Charles's directive had at last been complied with. And what is particularly relevant to this story of Greenwich time is that Maskelyne's almanac was based on the

*Author's italics.

Greenwich meridian. Up to that time, seamen usually expressed their longitude as a certain number of degrees and minutes (or leagues) east or west of their departure point or their destination – 3° 47′ west of the Lizard, for example. But now any navigator using Maskelyne's *Nautical Almanac* to find longitude astronomically – and a very high proportion of the world's deep-sea navigators began to do so from 1767 – must end up with an answer based on the Greenwich meridian. Indeed, from 1774 to 1788 this applied even to those using the official French almanac *Connaissance des Temps*, where, with Maskelyne's agreement and assistance, the British lunar-distance tables (based on Greenwich) were reprinted verbatim, despite the fact that all the other tables in the almanac were based on Paris. The navigator having obtained a Greenwich-based longitude, needed to plot his position on a chart. So map and chart publishers the world over began to provide longitude graduations based on Greenwich, to such an extent that, when the need eventually arose for a prime meridian for longitude and time to be agreed internationally, it was Greenwich that was chosen (rather than, say, Paris), largely because, by that time, no less than 72 per cent of all the world's shipping tonnage was using charts based on Greenwich. And it was the publication in 1766 of *The Nautical Almanac* which had started the chain of events described above.

The marine chronometer in Britain

In Maskelyne's 'Explanation' to the first *Nautical Almanac*, he told why tabulations were given in apparent solar time rather than mean solar time. But he ended up with the following paragraph: 'But if Watches made upon Mr John Harrison's or other equivalent Principles should be brought into Use at Sea, the apparent Time deduced from an Altitude of the Sun must be corrected by the Equation of Time, and the Mean Time found compared with that shewn by the Watch, the Difference will be the Longitude in Time from the Meridian by which the Watch was set; as near as the Going of the Watch can be depended upon.'[45]

The story of how John 'Longitude' Harrison (1693-1776) eventually received the 1714 Longitude Act's highest award has been told often and well.[46] Harrison, son of a country carpenter, was born near Wakefield in Yorkshire but moved with the family to Barrow, near Barton in Lincolnshire opposite the port of Hull at an early age. John began by following his father's profession but soon turned to clockmaking.

By 1727 he and his brother had made two very superior precision clocks incorporating many new ideas, the most enduring of which was the gridiron pendulum, an apparatus which remains the same

effective length whatever the temperature (the longer the pendulum the slower the beat and vice versa, so for good timekeeping it must not change length).

He visited London about 1730 to learn more of the enormous awards being offered by Parliament for a solution to the problem of longitude at sea, one approach to which lay in his own field of precision timekeeping. Harrison was introduced to the Astronomer Royal, Halley (who was a Commissioner for Longitude) and to George Graham (1673-1751), one-time partner to Thomas Tompion and the most influential clockmaker of his day. Graham saw Harrison's potential, lent him money, and persuaded both the East India Company and Charles Stanhope (son of the first Earl Stanhope, who had played a leading part in the passing of the Longitude Act twenty-five years before) to grant Harrison sums of money to help make his first sea clock.

Above. John Harrison, with his H3 timekeeper in its original case behind. Oil painting by Thomas King, c. 1765-6. Below. Harrison's first marine timekeeper H1, 1730–5 (left) and H2, 1737–9 (right), which is a more sophisticated version.

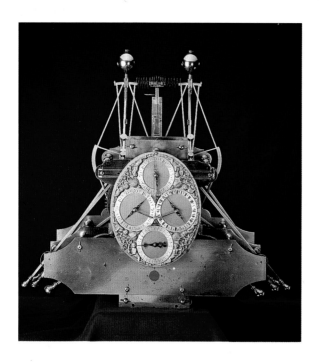

In recent years it has been the practice to call Harrison's marine timekeepers H1, H2, etc., a practice we will follow here. H1, a large and heavy machine three feet high, was completed in 1735. Thanks to the influence of Halley and Graham, the Board of Longitude arranged a sea trial: Harrison accompanied H1 to Lisbon and back in HM Ships *Centurion* and *Orford* in 1736. H1 performed well but Harrison was not entirely satisfied and, in any case, the Act stipulated a voyage to the West Indies to qualify for the higher awards. At a meeting of the Board of Longitude on 30 June 1737 – the first ever – the Commissioners granted Harrison £250 to make another 'sea-clock', with a promise of more money if the clock was successful. H2 was completed in 1739 but Harrison was not satisfied and proposed to make a third machine. Despite slow progress – H3 took nineteen years to complete – the Board and the Royal Society kept faith in Harrison, the former granting him a total of £3,000 between 1741 and 1762, the latter awarding him in 1749 their highest award, the Copley Medal, as 'the author of the most

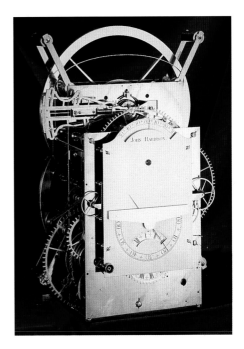

Above. H3, 1740–59, which convinced Harrison that large sea timekeepers would never be accurate enough to establish longitude. It incorporates two innovations: the bimetallic strip (for temperature compensation) and the caged roller bearing. Below, left and right. The silver-cased H4, 1755–9, was the prototype, 13 cm (5 in) across, that finally won Harrison the prize.

important scientific discovery or contribution to science by experiment or otherwise'.

In July 1760, while the Seven Years War was in progress, Harrison declared H3 ready for a trial to the West Indies under the conditions of the 1714 Act. But he also produced at the meeting a large watch, known today as H4, which, he said, 'answered beyond his expectation.'[47] The Board gave Harrison another £500, making a grand total of £3,250 since 1737, to finish adjusting H4, so that both H3 and H4 could be tried together. In the event, Harrison decided to stake his all on the watch which, on 18 November 1761, was embarked with his son William (John was 68) in HMS *Deptford*, bound for Jamaica. They reached Portsmouth on return in March 1762.

Though Harrison claimed H4 had more than complied with the provisions of the 1714 Act, the Board thought otherwise, awarded him £1,500 with a promise of £1,000 later, and said there would have to be another trial. On 8 April 1763, following a petition from Harrison, an Act of Parliament received the Royal Assent promising £5,000 if he would disclose details of the watch to a committee of experts appointed by Parliament.[48] Indeed, as we shall see, the Académie des Sciences in Paris sent representatives to attend these disclosures, despite the fact that France was still technically an enemy state. (The Treaty of Paris was proclaimed on 22 March and ratified on 5 May 1763.) Maskelyne's *British Mariners' Guide*, advocating the rival lunar-distance method, was published the same year. Harrison saw this as a threat: someone – perhaps even Maskelyne himself – might win the main prize before he, Harrison, was ready with his alternative timekeeper method. He therefore decided to abandon the £5,000 offered by Parliament and refused to disclose his secrets, but instead pressed for the second West Indies trial to settle the matter of the £20,000 reward once and for all.

The next trial took place in HMS *Tartar*, which sailed from Portsmouth for Barbados on 28 March 1764, with William Harrison and H4 embarked. Great precautions were taken to make the trial a fair one: Maskelyne was sent ahead to settle the longitude of Barbados by observations ashore; and a particularly accurate clock by John Shelton (the same who made the clock taken to Barbados by Maskelyne) was borrowed at Harrison's suggestion from the third Duke of Richmond (great-grandson of Charles II and the Duchess of Portsmouth) for use in observations at Portsmouth before the voyage.[49]

The story of Harrison's fight to get his hoped-for reward is too long to tell in full. Although on the face of it H4's latest results satisfied the 1714 Act, the Board was still not prepared to recommend the award of the full £20,000 without further proof that H4's results were not a matter of chance. In May 1765 Parliament passed an Act which effectively changed the rules of the game: Harrison would get £10,000 as soon as he disclosed his secrets and handed over all longitude machines to the Astronomer Royal – by then Maskelyne himself; but the second £10,000 would only be awarded 'when other timekeepers of the same kind shall be made' – and proved to be accurate enough to find the longitude to 30 miles.[50] The same Act authorized the awards to Mayer's widow and to Euler, and directed the Board to compile and publish the *Nautical Almanac*. Harrison was in despair. He was by then 72, his eyesight and general health were failing. However, he complied – under protest. H4 was put on trial at Greenwich and was then handed over to a London watchmaker, Larcum Kendall, to be copied, while Harrison and his son started making another longitude watch, H5, which they completed five years later.

The story of Kendall's copy, K1, during Captain Cook's second voyage (the most important secondary aim of which was the testing of K1 and three chronometers by John Arnold (1736-99), Harrison having refused to send H5) is remarkable.[51] In a voyage lasting almost exactly three years, in the Antarctic as well as the Tropics, it performed magnificently, the uncertainty of its daily rate never exceeding 8 seconds (2 nautical miles on the equator). When the voyage was nearly over, Cook wrote to the Secretary of the Admiralty from the Cape: 'Mr Kendall's Watch has exceeded the expectations of its most zealous Advocate and by being now and then corrected by lunar observations [Cook had the *Nautical Almanac*] has been *our faithful guide through all the vicissitudes of climates.*'[52] Perhaps even more telling are the comments of William Wales, the Board of Longitude's astronomer who sailed in the *Resolution* with Cook on this same voyage: 'From the preceding account it appears to what an amazing degree of accuracy the ingenious Inventor of this watch had brought this branch of mechanics so long ago as the year 1762, or 3; and at the same time what room is yet left for future improvements by other Artists: but let no man boast that he has excelled him, until his machines have undergone as rigorous a trial as this has done.'[53]

In the meantime Harrison had taken the bold step of approaching King George III personally. As a result, the King arranged for H5 to be put on trial at his private observatory at Richmond, today called Kew Observatory. When the whole story was retailed to him the King is reported to have said: 'By God! Harrison, I will see you righted!' Finally, the Act 13 Geo.III *c* 77 was passed in June 1773, granting the 80-year-old John Harrison £8,750, just £1,250 less than he had hoped for. (Over the years, he received a total of £23,065 under the Act of 1714, but this included expenses.)

*Author's italics

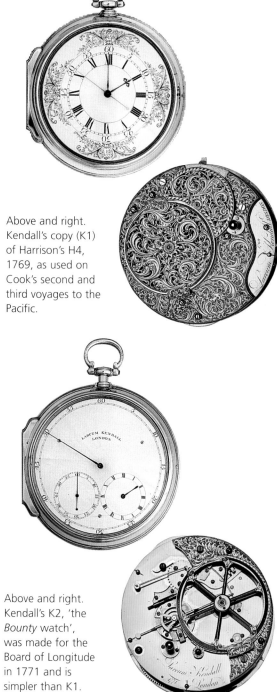

Above. Captain James Cook, painted by Nathaniel Dance in 1775, between his second and third voyages to the South Seas. Below. John Adams, last survivor of the *Bounty* mutineers. Kendall's K2 (lower right), was issued to Captain Bligh for his ill-fated voyage of 1787–9, and taken to Pitcairn Island by the mutineers. Adams sold it to Captain Folger of the American whaler *Topaz*, which found him there in 1808.

Above and right. Kendall's copy (K1) of Harrison's H4, 1769, as used on Cook's second and third voyages to the Pacific.

Above and right. Kendall's K2, 'the *Bounty* watch', was made for the Board of Longitude in 1771 and is simpler than K1.

John Harrison died on 24 March 1776, just eight months after Captain Cook had returned from a voyage which proved beyond doubt that it was possible to make a satisfactory longitude watch. But those that Harrison produced were experimental and expensive. It was left to younger colleagues to design chronometers (as they started to be called after 1780) cheap enough for the ordinary navigator. In the early days the names of John Arnold and Thomas Earnshaw were important. They sold chronometers for 60 guineas or so, whereas Kendall had received £500 for his copy of H4; both received rewards from the Board of Longitude for their developments in the design of chronometers.

John Harrison proved the whole concept possible and demonstrated the scale and fundamental principles upon which the longitude timekeepers should be made. But it was John Arnold who showed the world that what Harrison had proposed was practicable, on a simple and relatively inexpensive scale, and that such timekeepers could be made in large numbers. To Earnshaw must go the honour of devising manufacturing methods – coupled with rigid quality control – that might almost be called mass production. Earnshaw made chronometers that were used in ships of all nations for fifty years or more and it was his design that would, with little improvement, be made by artisans of many nations for over a hundred years.[54]

The British East India Company was early in insisting that all its ships should carry chronometers. The Royal Navy was somewhat slower: it was 1840 or later before ships carried chronometers in home waters, for example. Nevertheless, the British Admiralty continued to stimulate technical development, particularly by instituting annual chronometer trials which took place at Greenwich from 1821, with prizes for the best chronometers submitted.[55] From the very first it was to Greenwich time that those who used the *Nautical Almanac* set their chronometers.

The marine chronometer in France

But it was not only in Britain that the marine chronometer was being developed. In France two names stand out in this connection: Pierre Le Roy (1717-85), who succeeded his famous father as clockmaker to the King; and Ferdinand Berthoud (1729-1807), who was born in Switzerland but who spent most of his working life in France.[56] In 1754 Le Roy and Berthoud each deposited with the Académie descriptions of their respective marine timekeepers. Le Roy completed his first in 1756 but it was never properly tested. Berthoud's No.1 was ready in 1763: it was tested ashore by the astronomer Charles-Etienne Camus when it showed variations up to 16 minutes in 24 hours. In 1763 also, Le Roy presented his second, three feet high, but this received no better trial than his first.

As we have seen, the British Parliament the same year promised John Harrison £5,000 if he would disclose details of H4 to a committee of experts, in an Act which received the Royal Assent on 8 April 1763. On 21 March, four days before the relevant Bill had been considered by a Committee of the House of Commons and nine days before it was passed by the Commons to the Lords, the Duc de Nivernois, the French Ambassador, wrote to his superior in Paris, the Duc de Praslin, saying that he had been told that the examination would take place in public (which in fact was not the intention) and that he had been charged to ask 'whether you would wish to send a Frenchman here to be a witness and part of the examination. . . It is from Mr. Mackensie that I have all these details and his brother who is a great connoisseur and protector of the arts who interests himself greatly in these events.'[57] The informant was actually James Stuart Mackenzie, younger brother of the 3rd Earl of Bute who was then Prime Minister (though he was to resign on 18 April); both brothers were distinguished amateur astronomers. Nivernois's letter was passed to the Académie who submitted the names of Camus and Berthoud, to be joined by Joseph-Jérôme Lefrançais de Lalande, astronomer and editor of the French almanac *Connaissance des Temps*, who happened to be already in London on a private visit. Lalande was a personal friend of many of those connected with the Harrison affair and it has been suggested that his presence in London at the material time had some influence on the apparent issuing of the invitation.[58]

Camus and Berthoud joined Lalande in London on 1 May. Soon after, Mackenzie – who had been appointed Lord Privy Seal for Scotland on 15 April, just before his brother's resignation as Prime Minister – told Nivernois that he could no longer mix in the Harrison affair. On 8 May the French 'Commissioners' were taken to meet Harrison, when they were allowed to see H1, H2, and H3, but not H4. Lalande had seen H4 earlier, but only the outside. The Frenchmen waited a month in London, then grew impatient. On 2 June Camus wrote to the Earl of Morton, vice-president of the Royal Society and one of the commissioners appointed by Parliament to examine the Harrison timekeeper, saying that the French King had been assured that the secrets of Harrison's machine would be communicated to the French commissioners because the British Parliament wished all nations to profit from the discovery. In spite of these assurances, however, there was still no news of any plans for the disclosures to be made, though there was a rumour that nothing was to be done until Parliament reassembled. The French party, he reminded Morton, had already been in London a month but, with school examinations looming, he personally had to return to Paris by mid-June and could not there-

Le Roy's 'Montre marine A' (for *ancienne*), 1766, preserved at the Conservatoire National des Arts et Métiers, in Paris.

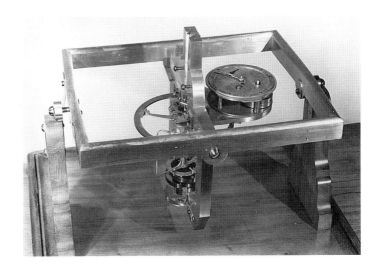

fore wait for Parliament to reassemble. Could Morton let him, Camus, know when the King and the Académie should send French commissioners to London again to attend disclosure of the secrets?

Morton, who seems to have been acting as Chairman of the British commissioners, sent a reply the following day. 'I am altogether a stranger to any invitation or the repeated assurances which you allege were given; . . . at the same time, if Mr Harrison had thought proper to comply with the demands of the Commissioners, I for my part should have been extremely well pleased that you, Monsieur, together with Messrs De la Lande and Berthoud had favoured us with your presence at the Examination of his Machine. . .' However, continued Morton, despite the fact that it was Parliament's intention that his machine should be made public for the common benefit of mankind, there was nevertheless no obligation upon Harrison to divulge his secret; also it was a matter of doubt whether Harrison would allow anyone not specifically named in the Act to be a witness.[59] The French delegation returned empty-handed to Paris shortly after Morton's letter; soon after that, Harrison made known his decision not to disclose his secrets for the time being. Although Le Roy stated later that the British Admiralty had invited French experts to witness Harrison's disclosure in 1763,[60] there is no mention of this in official British records so, in view of Morton's letter, it seems no official invitation was in fact extended, whatever the French Government may have believed at the time.

In France in 1764 Le Roy produced another marine timekeeper, half the size of his previous one, while Berthoud produced his Nos. 2 and 3, the latter being given a month's trial in *L'Hirondelle* corvette in October 1764 with very mediocre results. Early in 1766 Berthoud made another visit to England. Saying he had been studying English and knew all those in England concerned in the Harrison affair, Berthoud suggested to the Minister of Marine that another visit to learn the secrets of H4 would be worth while. He had been in touch with Short who had said that Harrison would co-operate if offered a reward of £4,000. In the event, the Minister of Marine offered no more than £500, which Harrison scornfully turned down as *'une si petite bagatelle'*.[61] But Berthoud did receive information from the watchmaker Thomas Mudge (1715-94), inventor of the lever escapement and one of those who attended the final disclosure of the secrets of H4 in August 1765. Mudge was later questioned by the Board about this breach of security which, as it turned out, did not matter much; Berthoud made little use of Harrison's inventions in the future.

Back in France the same year, Le Roy presented to King Louis XV his masterpiece – a wonderful marine timekeeper of completely original design which later came to be known as 'A' (for *ancienne).*

The following year, 'A' and a second timekeeper 'S' (for *seconde)* were tested at sea in a privately owned frigate *L'Aurore,* belonging to the Marquis de Courtanvaux, vice-president of the Académie Royale des Sciences. Results were good, but not good enough to qualify for the Prix Rouillé offered for that year 'to determine the best way of measuring time at sea . . .'[62] However, in 1768 the frigate *L'Enjouée,* with the astronomer Jacques-Dominique Cassini on board, made a voyage of 161 days to Newfoundland and back; 'A' and 'S' performed brilliantly, gaining for Le Roy a double Prix Rouillé' of 4,000 livres (about £170 sterling). The Académie decided to offer a double prize once again in 1773 for competition on the same subject – the best way to measure time at sea.

In the meantime Berthoud's Nos. 6 and 8 were being tested by Fleurieu and Pingré in the frigate *L'Isis,* in a twelve-month voyage in the Atlantic. Results were mediocre but the King did grant Berthoud, who seems to have been appointed chronometer-maker to the Navy, a pension of 3,000 livres (£128). It was then decided that the chronometers of Le Roy and Berthoud ought to be tested alongside each other, so Verdun de la Crenne in the frigate *La More,* sailed from Brest in October 1771 with the navigator Borda and astronomers Pingré and Mersains, with Le Roy's marine timekeepers 'A', 'S' and a watch, *La petite ronde,* Berthoud's No. 8, and two by Arsandeaux and Biesta on board. The voyage to Spain, Newfoundland, the Arctic and Copenhagen lasted a year. Le Roy's 'S' and Berthoud's No. 8 gave good results and the former won for Le Roy the 1773 double prize of 4,000 livres, Berthoud declining to compete in view of his official position.[63]

This marks the end of the experimental period for French chronometers. Thereafter, they began to be used regularly in ships at sea, particularly in voyages of exploration by navigators such as La Pérouse, Dentrecasteaux, Baudin and others, who carried chronometers by Ferdinand Berthoud, by his nephew Louis Berthoud, and by the Englishman John Arnold. France was somewhat later than Britain in the manufacture of chronometers but quality was high and the names of Abraham Louis Breguet, his son Louis, and Henri Motel deserve special mention.

The foundation of the French Bureau des Longitudes

On 7 Messidor, year III (25 June 1795), people's representative Grégoire (later, Bishop of Blois) proposed to the National Convention that France should follow Britain's example by having a Board of Longitude. In a long speech he quoted Themistocles as saying that whoever was master of the seas was also master of the world. The English had proved this to be so, he said, particularly in the war of 1761. And because of it, she had become a great power whereas, by all the normal rules, she should play a merely secondary role in the political order. But, said Grégoire, British tyranny must be stifled. And what better way than by using the methods she herself adopted?

Britain realized that, without astronomy, there could be no commerce, no navy. So she had gone to incredible expense in pushing astronomy to the point of perfection. And it was to her Board of Longitude that much of the credit for this was due: not only did it have enormous sums of money to disburse, but it also published the *Nautical Almanac* – admittedly on a French model – which had become their seamen's handbook.[64]

Grégoire suggested the foundation of a Bureau des Longitudes, on the British model but smaller and more manageable, to superintend the activities of Paris Observatory and the observatory of the École Militaire, and to oversee the publication of the *Connaissance des Temps*, then, as now, under private proprietorship. So the Bureau was founded that same year, the founder members being the geometers Lagrange and Laplace; the astronomers Lalande, Cassini, Méchain and Delambre; the retired navigators Borda and Bougainville; the geographer Buache; and the instrument-maker Caroché, the professions of the members and their number having been laid down in the Law of 1795, which was passed immediately thanks to the eloquence of representative Grégoire. The French Bureau still survives and remains a most prestigious scientific body, superintending the Paris Observatory and the *Connaissance des Temps*, publishing an important *Annuaire*, and acting as the French Government's chief adviser on scientific matters concerned with navigation.[65]

Britain's Board of Longitude was wound up in 1828, as its original function, the discovery of the longitude, had been achieved. Nevertheless, in 1918 the British report on the Conference on Standard Time at Sea was so impressed with the achievements of the French Bureau that they recommended that the British Board of Longitude be re-established. But no action was taken.

Time-balls

As the name implies, a marine timekeeper is designed to *keep* time at sea. But for navigational purposes it is necessary to know the time in the first place, and the going of the timekeeper – the chronometer – must be checked periodically thereafter. In the early days of chronometers this could be done by lunar observations ashore or afloat (not very accurate), by stellar observations ashore with a sextant and artificial horizon, or by comparison with an observatory clock ashore. Whatever method was used, it was most unwise actually to move the chronometer (this might disturb its going), so a pocket watch had to be used as an intermediary, or a signal made from ashore which could be seen or heard on board. In the 1820s there are several reports of these signals being made from shore for the benefit of ships in harbour – a flag dipped, a gun fired, a searchlight eclipsed, a rocket fired. But these all seem to have been *ad hoc* arrangements; there were no regular time signals.

Captain Robert Wauchope, RN, seems to have been the first to propose that time-balls should be erected, in communications with the Admiralty in 1818 and again in 1824, when serving on the Cape of Good Hope station. In 1829, an experimental time-ball was set up at the entrance to Portsmouth Harbour, controlled by a visual signal from the Royal Naval College in the Dockyard, a little under a mile away. Then, in June 1833, he suggested to the Admiralty that one should be erected at Greenwich. This was passed on to John Pond, the Astronomer Royal, who took quick action.[66] In October, the following Notice to Mariners was issued:

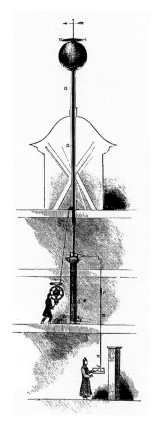

Admiralty, 28 October 1833.

The Lords Commissioners of the Admiralty hereby give notice, that a ball will henceforward be dropped, every day, from the top of a pole on the Eastern Turret of the Royal Observatory at Greenwich, at the moment of one o'clock P.M. mean solar time. By observing the first instant of its downward movement, all vessels in the adjacent reaches of the river as well as in most of the docks, will thereby have an opportunity of regulating and rating their chronometers.

The ball will be hoisted half-way up the pole, at five minutes before One o'clock, as a preparatory signal, and close up at two minutes before One.

By command of their Lordships

John Barrow [67]

The Greenwich time-ball of 1833, showing how the ball was hoisted to the top of the pole daily at 12.58 and released by an assistant standing in front of the clock at 1.00. From *The Illustrated London Almanack* for 1845, p. 28.

One o'clock was chosen for dropping the ball because, at noon, the astronomers might be busy *finding* the time.

The apparatus, constructed in 1833 by Messrs. Maudslay & Field at a cost of about £180, remains substantially unchanged today except that, since 1852, the moment of drop has been controlled by an electric current from a master clock (see next chapter) and, since 1960, the raising of the ball has also been made automatic.[68]

Wauchope also put his suggestion to the East India Company, resulting in time-balls coming into operation in Mauritius in 1833, St Helena in 1834, Cape of Good Hope in 1836, and Madras and Bombay in the 1840s. He had also written to the French and

Shepherd's mean solar standard 'motor clock' and the apparatus for dropping the Greenwich time-ball, as brought into use from 1852. From *The Graphic*, 8 August 1885.

Below, left. The Observatory at Sydney, Australia, was built specifically to provide a time service. The time-ball mechanism was made by Maudsley's of London and inaugurated in 1858.

Below, right. The restored 1876 time-ball station at Lyttelton, in New Zealand's South Island, now overlooks a modern container port, rather than the sailing clippers it first served. The building is one of the few that were specially designed for the purpose.

American governments in 1830 but the first American time-ball, on the roof of the US Naval Observatory in Washington, D.C., did not come into operation until 1845 or 1846.[69]

Not only did the Greenwich time-ball – said to be the world's first public time signal – give Greenwich time to ships in London's river and docks, but, for the first time, it made Greenwich time regularly available to those ashore who could see it.

Right and below. The Greenwich time-ball rises half way up its mast at 12.55 pm. It rises to the top at 12.58 pm ,then drops at exactly 1 pm. The astronomers chose 1 pm as the time to operate the signal, because at noon they were busy with observing the Sun as it passed the local meridian.

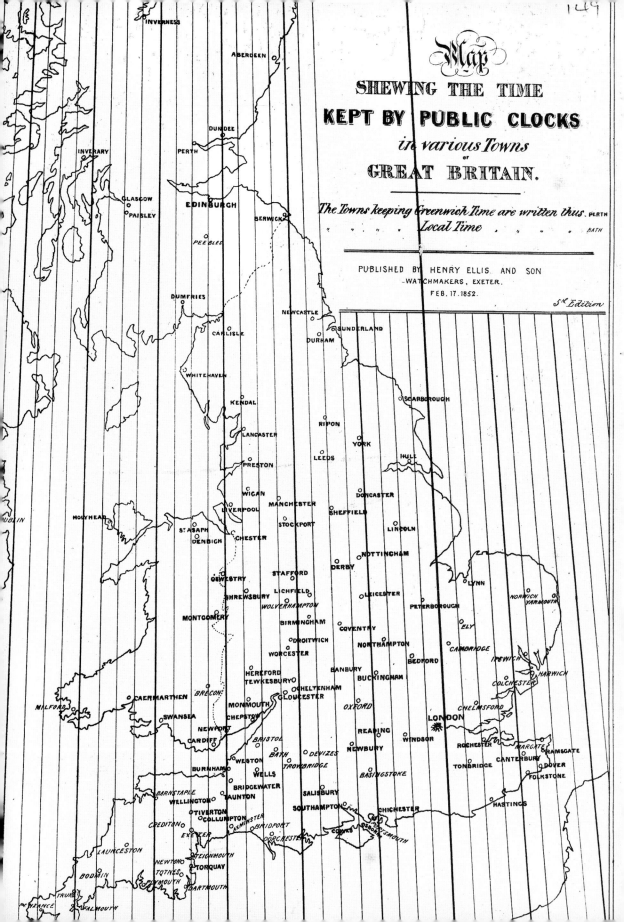

149

Map
SHEWING THE TIME
KEPT BY PUBLIC CLOCKS
in various Towns
of
GREAT BRITAIN.

The Towns keeping Greenwich Time are written thus . PERTH
" " " *Local Time* . " " . BATH

PUBLISHED BY HENRY ELLIS AND SON
WATCHMAKERS, EXETER.
FEB. 17. 1852.

5th Edition

FOUR

Greenwich time for Great Britain

1825 - 1880

S o far we have been considering time from the points of view of two very limited and specialist classes of user – astronomers and navigators. In this chapter we shall start considering time as it affects people in general – time for civil purposes. From the earliest period man has regulated his activities by the Sun, as indeed have animals and plants, by the daily alternation of day and night, and the yearly alternation of winter and summer. It is the first of these which concerns us here – the daily alternation of light and darkness, the time to work and the time to rest.

At first, the simple division of night and day sufficed for measuring periods of time shorter than a month (or 'moon'). To express a period of twenty-four hours some civilizations used the word *day* (as we do in the West today), some the word *night* (as did the Jews and Arabs, from which usage came the English word *fortnight*, for example). But soon this primary division was found to be inadequate and the day began to be divided into hours, sometimes, as we have already seen, *unequal hours* based on daylight and darkness, sometimes equal hours which divided up the whole day – from sunrise to sunrise, from sunset to sunset, from midday to midday, or from midnight to midnight – into periods of equal length.[1] At a later stage it became necessary to divide the hour, first into simple halves and quarters, later still into minutes *(prima minuta)* and seconds *(secunda minuta)*.

Early sundials consisted of a vertical rod or pillar – a gnomon – which sufficed to measure unequal hours. Sundials proper, capable of measuring equal hours, were in use in Athens at least from the time of Pericles. But, as we have already seen, sundial time has a disadvantage which only became apparent with the development of clocks and watches. Because the Earth's orbit around the Sun is not circular, and because of the tilt of the Earth's axis (which causes summer and winter), the time as shown on a sundial – *apparent* solar time – is not uniform, varying by as much as 16 minutes on either side of the mean during the year, the amount on any particular day being known as the Equation of Time (see diagram on p. 88).

Opposite. The towns in italic were still keeping local time in February 1852. From the edition of a map published by Henry Ellis & Son, Exeter, in February 1852.

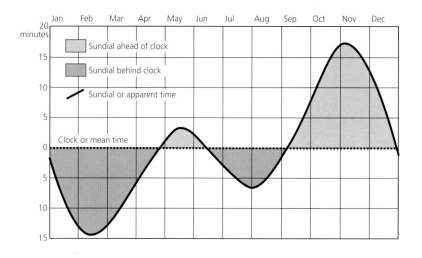

The Equation of Time.

This inequality of the Sun's motion in the ecliptic – that it seems to travel faster at some seasons than others – had been known since the 5th century BC when it had been discovered with the aid of the gnomon that the equinoxes and the solstices did not divide the year into four equal parts as might have been expected. But to the ordinary citizen this non-uniformity of sundial time was at first no great inconvenience. Until well on into the eighteenth century clocks and watches (which, within their limits of accuracy, showed mean solar time) were mostly for the rich and, in any case, their inaccuracy made the discrepancy between clock and sundial less obvious. However, by the end of the eighteenth century clocks and watches became commonplace and their performance improved greatly. Communities began to keep mean time in preference to apparent time – Geneva from 1780, England from 1792, Berlin from 1810, Paris from 1816. Arago recounts how, before the introduction of this change in Paris, the Prefect asked for guarantees from the Bureau des Longitudes (who had suggested the change) that there would be no trouble from the people at large. Would they not be worried that *midi* might not in future always occur in the middle of the day – which would be a contradiction in terms? In the event, the change occurred with almost no comment at all.[2] But, except for special purposes, the time kept was still local time, based on the meridian of the place concerned. That this involved each community keeping different times – the local time in London and Plymouth, for example, differs by some 16 minutes – was also no great inconvenience, because travellers were few and the rate of travel slow.

The coming of the railways from 1825 onwards, however, brought a new situation. As far as the railways were concerned, it was extremely inconvenient, to say the least, that there should be no standard time throughout the system. Even before the railways, this inconvenience made itself felt in at least one kind of land travel, the running of mail coaches strictly to timetable, first suggested in England by John Palmer, manager of the Bath Theatre, in 1782.

The service started in 1784 but, by 1792, the only coaches on the road were those supplied by one Besant. In addition to carrying His Majesty's Mails, the coaches were constructed to carry five passengers, four inside and one out. The coachman was a servant of the company, the mail guard a Post Office employee; both wore uniform. The guard was not only responsible for the safety of the mails in his charge but had also to see that the coach kept to time. He carried firearms and also a timepiece, regulated, it is said, 'to gain about 15 minutes in 24 hours, so that, when travelling eastwards, it might accord with real time. Of course, in the opposite direction a corresponding allowance was made.'[3] The mail coach tradition – that the person responsible for punctuality should carry an official timepiece – was continued by the railways, though without the allowances for easting and westing.

The lack of any system of uniform time was an inconvenience for astronomers also. The eminent Sir John Herschel devised Equinoctial Time, first described in a supplement to the *Nautical Almanac* for 1828, based on his Equinoctial Year which, by definition, started at the moment of the vernal equinox (when the Sun is vertically overhead on the equator, on 21 March each year give or take a day). Equinoctial time, a time-scale of astronomical origin but independent of terrestrial longitude, was thus the precursor of Ephemeris Time which was introduced for much the same reasons in 1952: the former seems seldom to have been used and the *Nautical Almanac* ceased giving data from which it could be obtained in 1876.

The period from 1820 to 1850 saw far-reaching developments in communications which were to change people's attitudes towards time and timekeeping: the first public passenger train in 1825; the first Atlantic crossing under steam power in 1827; Wheatstone's electric telegraph of 1836; mail sent by rail from 1838; Bradshaw's railway timetables of 1839; Bain's electric clock of 1841; the first public telegraph of 1843 (running alongside the Great Western Railway line from Paddington to Slough). By the 1840s there were at least three types of organization for which it was a grave inconvenience, to say the least, that different communities should keep different times – the Post Office, the railways and the telegraph companies.

Time distribution by hand

Luckily, as the need for a standard time became pressing, the means of satisfying that need became available – and, in Britain at least, the person to organize it was standing ready. The means was 'galvanism' (as electricity was then known), using the newly invented electric clock and electric telegraph; the person was George Airy (1801-92), seventh Astronomer Royal, one of the few scientists in England then in government service. But at first, means other than galvanism were used to distribute time. We have already heard about the Greenwich time-ball of 1833. In June 1856, the year the London and Greenwich Railway was opened, one of the assistants at Greenwich Observatory, John Henry Belville, started a weekly call on the principal chronometer-makers in London, taking with him a pocket chronometer that was set to Greenwich time.

The origins of John Henry Belville are rather obscure. According to notes left by Ruth Belville, his daughter by his third wife, there is little doubt that his parents were French. The notes record that he arrived in England with his mother in 1794, where they were befriended by the Countess of Bath (*sic*).[4] The only firm record, in the parish registers of St Pancras Old Church, states, however, that John Henry Belville was baptized there on 16 August 1794, giving a date of birth of 21 July; the most plausible explanation must therefore be that his mother, probably widowed, was pregnant when she arrived in England. It is also known that John Pond (1767-1836), who became Astronomer Royal in 1811, took great interest in his welfare, and that he moved to Greenwich, with Pond, on the day the latter took up his appointment. He was then sixteen years old, and it must be inferred that Pond was in some way involved in his education .

Five years later, in 1816, he was appointed Second Assistant at the Royal Observatory, where he was always known as Mr John Henry, it being thought expedient to disguise his French ancestry. In the same year Pond's new 10ft transit telescope was installed, and John Henry was engaged in working on this instrument. Subsequently he became involved in the observatory's work with marine chronometers, and with the dissemination of time. He was the first assistant to be in charge of Airy's great transit circle, from 1851 until his death in 1856. Airy paid tribute to him in his annual report the next year – 'the last who remained from Mr Pond's establishment, and one of the most

Airy's transit circle, which has defined the Greenwich meridian since 1851; this became the Prime Meridian of the world in 1884. From *The Graphic*, 11 Dec. 1880.

faithful and zealous of my coadjutors'[5] – which, from Airy, was praise indeed!

On Henry's death the time distribution service was continued by his widow who retired in 1892, to be succeeded by her daughter Ruth Belville, who continued the service until the 1930s. In her later years 'the Greenwich time lady' had some forty to fifty houses where she called once a week with her famous Arnold chronometer, conveying Greenwich time. She visited Greenwich Observatory each Monday morning when the chronometer was checked for her and a certificate was issued showing its error. The same watch was used throughout, a large-size silver pocket chronometer, No. 485/786 by John Arnold and Son, made for one of George III's sons but rejected because it was 'like a warming-pan'. It was subsequently selected after competition at the Observatory for the office it has since filled. It is now in the collection of the Worshipful Company of Clockmakers, to whom it was bequeathed by Miss Belville on her death, aged 90, on 7 December 1943.

Miss Ruth Belville, the 'Greenwich-time lady', who called at the Observatory every Monday, until the 1930s, to check her pocket chronometer, and then paid a weekly call on all the principal chronometer-makers in London.

Post Office time and railway time

In 1840 Captain Basil Hall, RN (1788-1844), explorer and one-time Commissioner for Longitude, wrote to Rowland Hill (1795-1879), author of the penny post and the adhesive postal stamp, and then an officer in the Treasury, suggesting that all post-office clocks throughout the different counties should be kept to London time.

> He proposed to regulate all post-office clocks in the Kingdom, by means of the time brought from London by the mail-coach chronometers; and he had no doubt that, ere long, all the town clocks, and, eventually, all the clocks and watches of private persons, would fall into the same course of regulation; so that only one expression of time would prevail over the country, and every clock and watch indicate by its hands the same hour and minute at the same moment of absolute time.[6]

So ran the report under the headline, 'Important to Railway Travellers. Uniformity of Clocks throughout Great Britain', in the first-ever number of *The Illustrated London News* in 1842, a report of a talk to the Birmingham Philosophical Institute by Abraham Follett Osler (1808-1903), a distinguished Birmingham meteorologist and businessman whose self-registering anemometer and pluviometer (measuring wind speed and direction, and the amount of rainfall) had been installed at Greenwich the previous year. Osler, demanding government action to institute 'British Time' throughout the kingdom, quoted Basil Hall who had, in turn, credited the late Dr. William Hyde Wollaston (1766-1828) with the original idea.

In fact, it was not the Post Office but the railways that eventually forced a uniform time on a not-unwilling population. In November 1840 the Great Western Railway ordered that London time should be kept at all its stations and in its timetables;[7] many other railways followed suit in the next few years, including the Midland and the South Eastern. In 1845 the Liverpool & Manchester Company petitioned Parliament to grant 'uniformity of time for all ordinary and commercial purposes throughout the land'. The petition was unsuccessful, but in January 1846 the recently constituted North Western Railway introduced London time at their Manchester and Liverpool termini, before which trains going up to London had observed Liverpool time while down trains observed London time in fact, it was Greenwich time.[8] In November the same year H. P. Bruyères, their General Manager at Euston, received a report about the late running of a train which was attributed to the fact that London time was kept on the line between Rugby and York (Midland Railway) whereas local Rugby time was kept at Rugby Station (run by the North Western) – and would the North Western change to London time, please?[9]

William Powell Frith's famous painting of 1862 shows a main-line departure of the Great Western Railway from Paddington Station, London.

In 1847 Henry Booth, secretary of the Liverpool & Manchester, published a broadsheet addressed to the Rt. Hon. Edward Strutt, Chairman of the Railway Commissioners, asking him to use his influence to get government action. According to Booth, the Post Office had already accepted Basil Hall's suggestions: '. . . accordingly, all their movements are regulated by "London time". By it is the great scheme of intercommunication adjusted, from one end of the kingdom to the other; the ever-varying longitude of a thousand post-towns is made subservient to the metropolitan chime of St Martin's-le-Grand.'[10] He pointed out all the anomalies which were beginning to occur with the population as a whole keeping a different time from the railways and telegraphs: the missed trains; Bradshaw's

A railway map of Great Britain in 1854, the decorative title to *Bradshaw's Railway Guide*, which appeared monthly from 1841.

Brunel conceived his transatlantic steamers as an extension of the Great Western Railway Company, from Bristol to New York. The SS *Great Western* conducted a regular service from 1838. Steam enabled railways and ships to be co-ordinated according to schedules which needed common timekeeping. Painting by Joseph Walter.

timetable which, being in local time, seemed to make east-west travel faster than when going from west to east; the mail that left Holyhead at midnight on Wednesday by Holyhead time, which happened to be Thursday morning by London time; the baby born in London early on Saturday, the news of whose birth could be received in Dublin by telegraph on Friday night. Booth asked that the proposed uniformity of time be authorized by Act of Parliament, commencing with the year 1848. He added darkly: 'I am not sanguine that the change recommended will be without opposition from the Astronomer Royal, or the Hydrographer's office.'[11] And so it proved: some thirty years would pass before there was any government action on the matter of legal time.

The railways, however, had no such inhibitions. On 22 September 1847 the Railway Clearing House (a body set up in 1842 to co-ordinate many aspects of railway operation in Great Britain) resolved 'that it be recommended to each Company to adopt Greenwich time at all their stations as soon as the Post Office permits them to do so.'[12] (The difference between London, St Paul's, and Greenwich time is 23 seconds.) On 1 December 1847 the London and North Western and the newly completed Caledonian Railway adopted London time 'in consequence of instructions received from the General Post Office'.[13] It seems likely that other railways in Britain conformed on the same date because, in *Bradshaw's Railway Guide* for January 1848, the London & South Western, London & North Western, Midland, Chester & Birkenhead, Lancaster & Carlisle, East Lancashire, and the York & North Midland Railways are all listed as keeping Greenwich time. We know from other sources that the Great Western, South Eastern and the Caledonian were doing likewise. However, in December 1848 the Chester & Holyhead earned the displeasure of *The Illustrated London News* which reported that the Directors had ordered 'that the clocks at all stations shall be regulated by the celebrated Craig-y-Don gun, which is 16 min. and 30 sec. after Greenwich time'. It added: 'This cannot fail to prove a great inconvenience to travellers.'[14] The Irish Mail from London to Holyhead, however, ran to London time from its inception in 1848. Each morning an Admiralty messenger carried a watch bearing the correct London time which, he gave to the guard of the Mail at Euston, thus maintaining the old mail-coach tradition. On arrival at Holyhead the watch was handed to officials on the Kingston boat, who carried it to Dublin. On the return, the watch was carried back to London and handed back to the Admiralty messenger who met the train.[15] This practice was continued until 1939.

The start of the Greenwich time service

With the Post Office and the railways keeping London time, many towns and cities followed suit with their public clocks. Manchester and Salford, for example, did so from 1 December 1857.[16]

Whatever the Astronomer Royal's private views may have been on whether a particular place should keep local time or Greenwich time, he very often stated from 1850 onwards that in his opinion it was a primary duty of the national observatory to provide Greenwich time wherever and whenever it was needed. In his annual report to the Board of Visitors for June 1849 Airy had this to say in discussing what changes, if any, were needed in the work of the Royal Observatory:

> Another change will depend upon the use of galvanism; and, as a probable instance of the application of this agent, I may mention that, although no positive step has hitherto been taken, I fully expect in no long time to make the going of all the clocks in the Observatory depend on one original regulator.
>
> The same means will probably be employed to increase the general utility of the Observatory, by the extensive dissemination throughout the Kingdom of accurate time-signals, moved by an original clock at the Royal Observatory; and I have already entered into correspondence with the authorities of the South Eastern Railway (whose line of galvanic communication will shortly pass within nine furlongs of the Observatory) in reference to this subject . . .[17]

Airy's main correspondent in the railway was Charles V. Walker, Telegraph Superintendent of the SE Railway Co., the first extant letter to whom is dated 19 May 1849[18] though there had been some earlier contact. On 26 May Airy sent his detailed proposals.

Things moved slowly, but 1851 brought significant developments. At the Great Exhibition in London's Hyde Park the public clocks were driven electrically – not too successfully as it turned out – according to the 1849 patent of Charles Shepherd, of 53 Leadenhall Street.

The second development concerned the laying of a submarine telegraph cable from Dover to Calais. After an unsuccessful attempt in August 1850, a cable was successfully laid across the Channel on 25 September 1851, the news of which reached assembled scientists at the Great Exhibition just as Queen Victoria was leaving the platform after formally declaring the exhibition closed.[19] The Channel cable gave an additional reason for Airy to pursue his plans for the 'galvanic connection' with the SE Railway: if Greenwich and Paris Observatories could be directly connected by telegraph, it would be possible to find the difference of longitude very accurately because observations in Paris could be registered on recording surfaces at Greenwich, and vice versa.

PUNCH'S FANCY PORTRAITS.—No. 134.

SIR GEORGE B. AIRY, K.C.B., F.R.S.,

THE ASTRONOMER-ROYAL WHO DESERVED THE GRATITUDE OF HIS COUNTRY
FOR HAVING "CORRECTED THE ATMOSPHERIC CHROMATIC DISPERSION."

Sir George B. Airy, KCB, FRS, caricatured as the Greenwich time-ball. From *Punch*, 1883, p. 214.

Briefly, Airy's plan was this: an electric clock should be installed at the Observatory, so fitted that it would give electrical impulses *(a) every second* for driving 'sympathetic', or slave, clocks in the Observatory and elsewhere, and *(b) every hour* for dissemination of time signals along telegraph lines from the Observatory to Lewisham Station and thence along the normal railway telegraph lines to London Bridge Station. From there, time signals could be sent down the lines of the South Eastern Railway (SER) and also to the Central Telegraph Station of the Electric Telegraph Company (ETC) in Lothbury in the City of London for further distribution all over the country – to other railways, to post offices, to public clocks, and, via the submarine cable, to the Continent. Except for a minute or so every hour, the telegraph lines from London would be used for ordinary messages, but just before the time signal, Greenwich would be put in direct communication with, say, Edinburgh and Plymouth, via London Bridge and the ETC. The clock at Greenwich, called at first the Normal or Motor Clock but later the Mean Solar Standard Clock, had the additional facility that it could be put right electrically. The time signal itself was a simple electrical impulse, which could be made to ring a bell, drop a time-ball, fire a gun, cause a galvanometer to kick, operate a relay, light a light, or even put another clock right. And all this automatically, with the Greenwich clock actually pressing the button that made a gun fire instantaneously in, say, Newcastle, 280 miles away.

In September 1851 Airy began implementing his plans in earnest. Early in the month he had written to his opposite number at Paris Observatory; on the 19th he prepared a Draft of Agreement with the SER; on 6 October he started formal negotiations with the ETC; on 7 October he wrote to Shepherd asking for proposals for a suitable clock; on 26 November he asked the Admiralty for funds (£350 or less),

> for immediately effecting a galvanic connexion between the Royal Observatory Greenwich and the London Bridge Railway Terminus, for the three purposes:
>
> 1st Of regulating the principal clocks of London (the Royal Exchange clock and the clock shortly to be constructed for the New Houses of Parliament);
>
> 2nd Of sending every day a time signal to every part of Britain which is reached by a line of Galvanic telegraph;
>
> 3rd Of communicating with the principal foreign observatories.[20]

Shepherd replied swiftly, sending a sketch and an estimate of £40 for the master clock and ball apparatus (for dropping the Greenwich time-ball automatically), with sympathetic clocks at £9 each. The Admiralty approved the necessary funds on 18 December and Airy sent a formal order on 19 December to Shepherd for 'One automatic clock (with face and works) as described in your letter and drawing of October 18. One clock with large dial to be seen by the Public, near the Observatory entrance, and three smaller clocks: all to be moved sympathetically with the automatic clock.' The total bill (of 29 September 1852) came to £244, the master clock itself being £70 and the wall-clock at the entrance £75, considerably over the original estimate.

The telegraph lines to Lewisham Station were completed on 17 February 1852. On 4 June Shepherd's clock was installed in the North Dome. On 16 July the time-ball was dropped electrically for the first time. On 2 August it was reported that the master clock was going, driving sympathetic clocks in the Chronometer Room, the Computing Room, the Dwelling House (Flamsteed House, which was Airy's own residence), and the Gate clock, which was set going

Greenwich Observatory about 1870, showing the time-ball and electrically-driven 24-hour gate clock, both controlled by Shepherd's 'motor clock'. The time of the photograph is a few minutes before 19.00 astronomical time, which was 7 a.m. civil time.

on the 14th. It was also providing the impulse to drop the time-ball at 1 p.m. daily.[21] In August all was ready for a wider distribution of Greenwich time. *The Times* of 23 August said:

> The arrangements for transmitting true Greenwich time automatically from the Royal Observatory by electric telegraph, and which have already been described in *The Times* [11 February 1852], are now completed and in practical operation on the South Eastern Railway.
>
> At noon and at 4 p.m. a single beat or deflection of the telegraph needle is visible at London, Tonbridge, Ashford, Folkestone and Dover, which represents Greenwich mean time. The first time signal from Greenwich was taken experimentally by Mr. C. V. Walker, in the clock-room at the London terminus, at 4 p.m., August 5, passing down to Dover. The 11 a.m. signal on August 9, was received at London in the presence of Dr. O'Shaughnessy, of Calcutta, and the noon signal of the same day in the presence of Mr. Herbert, the secretary of the South Eastern Railway Company. . .

In a letter to Airy of 20 August Walker said that for the Dover line – London, Redhill, Tonbridge, Ashford, Folkestone, and Dover – time signals would be taken at noon and 4 p.m. daily. For the North Kent line – London, Lewisham, Blackheath, Woolwich, Erith, Gravesend and Strood – signals would be taken at 2 p.m. The other twenty-one hourly signals would be available for the ETC at Lothbury for their own time-ball in the Strand (see below) and for distribution throughout the kingdom along the various other railway lines. In addition to the hourly signal, Shepherd's clock at Greenwich sent impulses every second to drive a sympathetic clock at London Bridge, which automatically made the various switching operations needed for sending out the hourly time signals.

The dial of Shepherd's Gate clock, the first to show Greenwich Time to the public, which it still does.

Procedure at the Royal Observatory

In essence, Airy's time distribution system at the Observatory itself was this:

(1) *Find the time* by astronomical observations of the so-called 'clock stars', using the transit circle (see Appendix II), generally at night – every night – when weather permits.

(2) *Correct the standard clock* to show the time just found (or estimated if observations were clouded out), daily, immediately before the principal time signals at 10 a.m. (1 p.m. Sundays).

(3) *Send out the time signal.* every hour, on the hour.

(4) *Repeat the procedure* the following day.

Though new clocks were brought into use and radio came to augment the telegraph as a means of distribution, Airy's system of 1852, based upon the idea that the rotating Earth is the fundamental timekeeper, remained virtually unchanged until the advent of the atomic clock in the 1960s.

Nationwide distribution of timesignals

At the start of the time service, all time signals went through the London Bridge switch-room and were sent from there (at the hours not needed by the SER itself) to the ETC's Central Telegraph Station at Lothbury in the City of London for distribution to other railways, post offices, etc. When the underground cables from the Observatory to Lewisham became defective in 1859, overhead lines were installed direct to Lothbury (and later to the London District Telegraph Co.'s headquarters as well) so that the telegraph companies' signals no longer had to go via the SER. But a line also went via Greenwich Station to the SER for the Deal time-ball (see below) at 1 p.m., for the British Horological Institute (for chronometer makers) at 2 p.m. and 8 p.m., and for the SER itself at all other hours, on the hour.[22] From London onwards, the time signals went to remote stations on telegraph lines used for normal messages and it was necessary to clear the lines of other traffic for a few minutes either side of the time signals (the principal ones being 10 a.m. and 1 p.m.), so that the signal from Shepherd's clock at Greenwich could pass direct to, say, Glasgow at precisely 10 a.m.

Single-needle telegraph instrument, commonly used in the nineteenth century for sending and receiving time signals. From *The Illustrated London News*, 28 Nov. 1874, p. 504.

At first, the necessary switching was done at Lothbury by a clock similar to that at London Bridge, though not controlled from Greenwich; but, about 1864, an apparatus to perform this switching automatically was designed by Cromwell F. Varley, electrician to the Electric and International Telegraph Company (E & ITC: the ETC and the International Telegraph Company had merged in 1855). Driven by an accurate pendulum clock which could be kept correct mechanically without having to touch the pendulum, the 'chronopher' (from χρονος, *time*; and φορος, *I bear*) placed the telegraph lines concerned in direct touch with Greenwich 1 minute 50 seconds before the time signal and put Greenwich out-of-circuit 1 minute 20 seconds after the hour. The original chronopher had two circuits, one serving stations in London, one provincial stations. In 1864 the local circuit sent *hourly* time signals to a clock and six bells within the Central Telegraph Office building; to time-balls at the E & ITC's office in the Strand and Bennett's of Cornhill (the City Observatory); to clocks at the General Post Office, Lombard Street post office and Dent's, the clock and chronometer maker in the Strand; and a signal to the Westminster clock ('Big Ben'). The provincial circuit sent *daily* time signals to telegraph offices in Manchester, Liverpool, Birmingham, Glasgow, Bristol, Portsmouth, Bath, Cardiff, Brighton, Hull, Derby and Lowestoft, to the royal residence at Sandringham, and to the London, Chatham and Dover Railway.[23] When the Central Telegraph Station moved to St Martin's-le-Grand in 1874, after the telegraphs were taken over by the General Post Office, a second and larger chronopher was added to the system.

The ETC's Strand time-ball

In 1852 the ETC's West End office was in 448 West Strand, a building (of which the façade survives) designed by John Nash with two 'pepper-pot' towers at each end, immediately opposite today's Charing Cross Station. Edwin Clark, ETC's chief engineer, wrote to Airy on 27 February explaining his plans for a time-ball on one of the 'pepper-pots', to be dropped at 1 p.m. daily by an impulse from the Greenwich clock. As *The Times* explained while it was being erected in June, there was to be a time-ball 'which is intended by means of sympathetic electrical action to fall every day simultaneously with the well-known ball on the top of the Greenwich Observatory, between which and the Strand the electric wires have been completed for the purpose, so as to indicate to all London and the vessels below bridge exact Greenwich time.'[24] The ball was 5 feet in diameter, made of zinc, painted black with a broad white band around it, the shaft being topped with a weather vane with 'ETC' on it. The whole cost £1,000. The automatic dropping of the

The time-ball in the Strand, London. From *The Illustrated London News*, 11 Sept. 1852, p. 205.

Strand time-ball started on 28 August. 'The public assemble in crowds and the chronometer-makers think it a great boon.'[25] However, there were troubles. On 29 August, with great crowds watching, it dropped 28 seconds late; on 1 November it dropped 1½ minutes early. Never a great success because of technical difficulties, it was discontinued after a few years.

On the railways, however, all seems to have gone well. On 30 October the following general order was passed to the SER:

Electric Telegraph, Tonbridge,
October 30th, 1852.

South Eastern Railway

General Order

GREENWICH MEAN TIME SIGNALS

The Astronomer Royal has erected Shepherd's Electro-Magnetic Clock at the Royal Observatory, for the transmission of Greenwich Mean Time to distant places.

On and after November 1st, the needle of your Instrument will move to make the letter N precisely at . . . o'clock every day. [Different stations received time-signals at different hours.]

Abstain from using the instrument for Two Minutes before that time.

Watch the arrival of the signal; and make a memorandum, for your own information, of the error of your Office Clock.

You are at liberty to allow local Clock and Watch Makers to have Greenwich time, providing such liberty shall not interfere with the Company's service and the essential privacy of Telegraph Offices, and the business connected therewith.

Engineer and Superintendent of Telegraphs

To Mr............................   ....Station.[26]

A similar order of the same date is known to have been passed to all stations of the Great Western Railway[27] and doubtless, through the ETC, to other railways as well.

The Deal time-ball

Writing to Airy on 12 January 1852 C. V. Walker said: 'Has it ever occurred to you to cause the Greenwich clock to drop a ball at Dover? It would not be impossible: it would be useful.'[28] On 13 April Commander Thomas Baldock wrote to the Admiralty suggesting that a time-ball should be dropped by galvanic current on one of the South Foreland lighthouses so that ships in the Downs, where those outward-bound waited at anchor for a favourable wind, could check the errors of their chronometers before taking departure.[29]

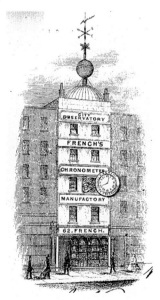

The time-ball in Cornhill, London, 1860. From a letterhead of J. French (late Bennett), chronometer maker, on whose premises it stood. Both this and the Strand time-ball were triggered by telegraph signal from Greenwich.

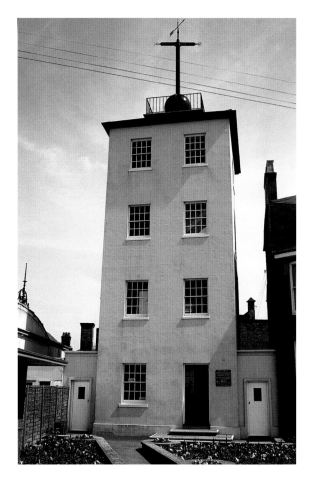

The time-ball at Deal, Kent, was put up on the former Navy semaphore tower there. It received a telegraph signal from Greenwich to drop the ball at 1.00 p.m., and served ships in the anchorage known as the Downs.

Opposite. The Nelson Monument, Calton Hill, Edinburgh. The time-ball was installed in 1854 by C. Piazzi Smyth, second Astronomer Royal for Scotland. Detail from a hand-coloured postcard.

In the event, it was decided that the ball should be erected in the old Navy Yard at Deal. Airy ordered a galvanic clock and ball apparatus from Shepherd on 12 November 1853 – £21 as per estimate. This came into operation on 1 January 1855, dropped by the 1 p.m. current from Greenwich, and continued until 1927. Soon after, a time-gun was installed at Dover Castle, fired at noon daily by the current from Greenwich.[30]

Airy recalled the institution of the system in a properly patriotic address to the British Horological Institute in the year 1865:

I can hardly say how the time signal system came to be first proposed, it was somehow, partly in conversation, partly in other ways, how, I cannot exactly say, but to Mr. C. V. Walker, Mr. Edwin Clark, Mr. Latimer Clark, and afterwards Mr. C. F. Varley, is the existence of the system mainly due.

The Deal time ball was not proposed by me, though I have taken great pains to render it efficient. I have indeed always considered it a very proper duty of the National Observatory to promote by utilitarian aid the dissemination of a knowledge of accurate time which is now really a matter of very great importance. The practical result of the system will be knowledged by all those who have travelled abroad. We can, on an English railway, always obtain correct time, but not so on a French or German railway, where the clocks are often found considerably in error.

Early developments in time-distribution

Developments in the first twenty years or so of the Greenwich Time Service, too numerous for each to be treated in detail, can be summarized as follows:[31]

1852 Airy's chronograph installed at Greenwich, automatically recording seconds impulses (later, two-seconds impulses) from the Sidereal Standard Clock on the same paper chart as the precise times of astronomical observations on transit circle. It did not come into full operation until 1854.

1855 Four clocks in London post offices were regulated by signals from Greenwich, sending return signals to ensure correct functioning. Writing in 1868, Airy said: 'In the clocks at the Lombard Street Post-Office, I some years ago arranged an apparatus by which at noon every day a galvanic current from this observatory seizes the second-hand of the principal clock and turns it round (if necessary) so as to make it point to 0^s. This apparatus has been long in action without a failure.'[32]

1855 (July) The Electric Telegraph Co. (1846) and the International Telegraph Co. (1852) merged to form the Electric and International Telegraph Co. (E & ITC).

1856-7 Time-balls controlled by signals from Greenwich erected at the E & ITC's office in Liverpool and at the premises of French (late Bennett), watch and clock-makers, in Cornhill in the City of London (with the rather grandiose title of the City Observatory); the same current dropped time-balls at Greenwich, the Strand, Cornhill, Deal and Liverpool.

1859 Underground telegraph lines from Observatory to Lewisham failed. Replaced by six overhead lines to Greenwich Station.

1860 Time superintendent's desk set up at Royal Observatory with instrumentation permitting all operations connected with time distribution to be controlled from one place.

1860 E & ITC's Central Telegraph Station moved from Lothbury to Telegraph Street.

1861 Time-gun controlled by Royal Observatory, Edinburgh, set up at Edinburgh Castle. Time-ball on Calton Hill in operation since 1858. Airy reported new time-balls at Glasgow, Calcutta, Sydney and Quebec. Portsmouth time-ball now on the roof of the Royal Naval College in dockyard.[33] Also Batavia (Jakarta) and Valparaiso.[34]

1862 Great clock at Westminster – popularly 'Big Ben' – installed. Greenwich sent hourly time signal, with return signal twice daily. The clock was in no way controlled from Greenwich.

1862 London District Telegraph Co. (LDTC; founded 1859) received Greenwich time signals for distribution in London, particularly to chronometer makers and jewellers.

1863 Time-guns established in old Norman keep at Newcastle, and at North Shields, fired at 1 p.m. by current from Greenwich via E & ITC (later via GPO).

c.1864 Varley's first chronopher installed in the new Central Telegraph Office in Telegraph Street in the City of London.

By 1865 Clock installed at factory of Warren de la Rue in Bunhill Row, London, who '. . . estimates the annual saving to his firm by having exact time and enforcing strict attendance on his work-people, at £300 (besides some saving of gas and coals not taken into account) which is an amount that would otherwise be entirely lost, and of this he is able to make a return to his work-people in the way of additional privileges as respects holidays'.[35]

1865 (28 Jan.) and 1866 (11 Jan.) Snow and gales brought down all overhead lines from Observatory – two years running.

The Chronopher Room at the Central Telegraph Office, General Post Office, St. Martin's-le-Grand, 1874. The old chronopher is behind the man, the new one to his left. From *The Illustrated London News,* 19 Dec. 1874, p.576.

1866-7 Underground telegraph lines laid from Observatory to Greenwich Station.

1870 (1 Jan.) All electric telegraph companies taken over by General Post Office, including the time signals distributed by E & ITC and LDTC.

1872 (29 July) Post Office circular confirming that Greenwich time was to be kept in all post offices.

1872 (July) Messrs. Arnold & Lewis, jewellers and watchmakers, of St Ann's Square, Manchester, set up an electric clock surmounted by a time-ball, dropped at 1 p.m. daily (except Sunday) by current from Greenwich.[36]

1874 (January) GPO's Telegraph Department moved from E & ITC's Telegraph Street office to a new building facing the GPO at St Martin's-le-Grand. New and more elaborate chronopher for sending time signals in sixty different directions installed for 10 a.m. signal, old chronopher being used for the 1 p.m. signal. The following account by Mr H. Eaton of Post Office Telegraphs shows time-distribution state in 1874:

The Greenwich current is received hourly. This hourly current is transmitted to 10 subscribers (mostly chronometer-makers) in London. The method of observing the current varies, and is fixed by the subscriber. In two cases, time-balls are dropped on the top of the buildings; in some other cases, model time-balls are placed in the windows; and others again use an electric bell; while two or three have a simple galvanometer, and observe from the deflexion of the needle.

The Westminster clock records its correctness and errors at Greenwich, as does also the clock at the Lombard Street Post Office.

The 10 a.m. current is most extensively used for the Provinces. It is transmitted automatically to 21 provincial towns in England (where there are subscribers), to Guernsey, Edinburgh, Glasgow, Dublin and Belfast. In addition to the automatic sender, a sound-signal is established in the Instrument Room here; when heard, a current is sent by the clerks to over 600 offices in direct communication with the Central Telegraph Office, including the principal railway termini. Many of these offices redistribute the time-signal to the offices radiating from them, so that practically from the 10 a.m. current from Greenwich most of the post-office and railway clocks in the Kingdom are regulated.

The 1 p.m. current is transmitted automatically to nine provincial towns, viz., Newcastle, Sunderland, Middlesboro', Kendal, Hull, Norwich, Stockton, Worcester and Nottingham. At the first four named, guns are fired; at the others, the current is observed by means of time-balls or galvanometers.

With regard to the 10 a.m. current, I should have said that there is no rule as to the method of observing; the subscribers use the form of apparatus most suitable to themselves. At the Telegraph Office the signal is recorded or observed on the telegraph instrument.[37]

1888 Astronomer Royal proposed to discontinue 10 a.m. time-signal. Post Office objected: 'We cannot dispense with the 10 o'clock current from Greenwich. . . Every office throughout the United Kingdom is supplied with this time current by a system which has taken many years to establish. . . But for what purpose was Greenwich Observatory established, if it was not for the production of accurate time for national and imperial objects; and what object is of more consequence to the Government than the distribution of accurate time throughout the three Kingdoms to every post office and railway station?. . .'[38] The Astronomer Royal relented and the 10 o'clock signal continued.

Railway time v. local time

But meanwhile the matter of a uniform time – of railway time *v.* local time – had been under debate in the country as a whole. Soon after A. F. Osler had given his lecture in 1842, he proposed establishing a standard clock for Birmingham, and he collected funds with which he procured one of the highest class, made by Dent, which was placed in front of the Philosophical Institution. In due course, when he had fully established the clock's accuracy in the public's estimation, on one Sunday morning he altered the clock from Birmingham time to Greenwich time without mentioning it to anyone, and, although the difference was remarked upon, the Church and private clocks and watches throughout the town were one by one adjusted to Greenwich mean time, though at that period the country generally was still keeping local time.[39]

The first excursion train ran in 1844 and the great increase in railway mileage in the late 1840s revolutionized the social life and habits of the country. In a leader on 12 January 1850 *The Times* said: '. . . there must be thousands of our readers, we are sure, who, in the last three years of their lives, have travelled more and seen more than in all their previous life taken together. Thirty years ago not one countryman in 100 had seen the metropolis. There is now scarcely one in the same number who has not spent his day there. Londoners go in swarms to Paris for half the sum, and in one-third of the time, which in the last reign it would have cost them to go to Liverpool . . .'[40]

But the event which caused the greatest increase in passenger traffic on the railways was the Great Exhibition of 1851 in Hyde Park, London, which resulted in travel in Britain on an unprecedented scale, *The Times* reporting over six million visitors (only 75,000 were foreigners) – and almost all of those travelled by train. As we have seen, almost all the major railways had adopted Greenwich time by the year 1847 and, like

This gold traveller's pocket watch, made by Benjamin Lewis Vulliamy of London in 1847, has two independently set minute hands, one for local time and one for Greenwich Mean Time. A number of provincial public clocks also had double minute hands.

Birmingham, many cities in the North and Midlands began to set their clocks to 'railway time'. In Scotland, in letters dated 30 November 1847, the Chairman of the Edinburgh and Glasgow Railway Company wrote as follows to the Lords Provost of these two cities:

> I am requested to bring under your notice and that of the Council over which you preside, the intention announced by the principal English Railway Companies to adopt Greenwich instead of local time on their lines, to which alteration the Corporations of Liverpool and Manchester have resolved those towns shall conform.
>
> The directors of the Edinburgh and Glasgow Railway believe that it would be very advantageous to adopt the same time and will gladly do so if the authorities of the two cities which their lines connect agree to this change.[41]

The respective town councils agreed almost without demur, prompted by a leading article in *The Scotsman* of 4 December supporting the idea and forecasting that such a change 'would not disturb the arrangements of business and domestic life more than the errors of our clocks do now'. The change took place on the night of Saturday 29 January 1848 (the date chosen by the Post Office authorities), Edinburgh, Glasgow, Greenock, Stirling and Perth all resetting their public clocks on the same date.[42] The change, which meant that clocks were put forward 12½ minutes in Edinburgh and 17 minutes in Glasgow, seems to have met with virtually no public opposition, the average citizen being quite content to regulate his life by whatever time the public clocks – or the factory hooter – gave him. Of course there were some letters to newspapers deploring the change – any change. For example, *Blackwood's Magazine* of March 1848 carried an article entitled 'Greenwich Time: "The time is out of joint – oh, cursed spite" *(Hamlet)*'. In eight long pages the anonymous author castigated the Edinburgh Town Council for usurping the power of the

The inside of the Vulliamy watch shown opposite, engraved with the names of several English and continental towns and their time differences from Greenwich.

Almighty. 'What in the name of whitebait have we to do with Greenwich more than with Timbuctoo, or Moscow, or Boston, or Astracan, or the capital of the Cannibal Islands? The great orb of day no doubt surveys all these places in turn, but he does not do so at the same moment, or minute, or hour. . . .' The article ends with a sentiment which has an almost modern ring about it: 'It would much conduce to the comfort of the lieges, if, instead of directing the course of the sun, you were to give occasional orders for a partial sweeping of the streets.'[43] Predictably, there were as many approving letters, pointing out that the local time previously kept was not in fact 'God's time' but a time based on a fictitious Mean Sun, chosen because the Sun itself was such a bad timekeeper: indeed, in some months of the year, the new railway time was closer to God's time than before.[44]

However, in the east and west of England, opposition continued. On 21 June 1851 there appeared in *Chambers Edinburgh Journal* (paradoxically published in London) a light-hearted article entitled 'Railway-time aggression'.[45] The anonymous author started by saying: 'There is an aggression far more insidious in its advances than the papal one [*CEJ* was a great supporter of the Established Church]. . . We are now, in many of our British towns and villages, to bend before the will of a vapour, and to hasten on its pace in obedience to the laws of a railway company! . . .'[46] He then went on to recount some of the things that were happening, not specifically because railway time was being introduced but because there were two different kinds of time being used simultaneously – the bride who arrived at the Church at railway time while her groom (and the pastor, organist and choir) arrived at local time; the ruination of a dinner party because of the different times kept by host and guests; the friction of town (railway time) and country (local time); the difference in church clocks between High Church (local) and Nonconformist (railway).

On 2 October 1851 'E.S.H.' started a spirited correspondence in *The Times:*

Electric telegraph and local time
Sir, – Contemporaneously with the advance of railroads, and the invention of the electric telegraph, the difference of time arising from the variation in longitude of places has been considered objectionable; and, for convenience' sake, an uniformity of time – that of Greenwich – has been adopted throughout the Kingdom, with exception of a few places in the west of England.

By reason of the submarine telegraph, England will now be brought into immediate communication with France and the greater portion of Europe. The question therefore arises, what meridian should be determined on for universal adoption?

Swiftly, on 7 October, there came a riposte from one 'Chronos' of Greenwich, who was thought by many to be the Astronomer Royal himself:

Local time

Sir, – Your correspondent 'E.S.H.', after remarking that 'a uniformity of time – that of Greenwich – has been adopted throughout the Kingdom, with exception of a few places in the West of England, the difference of time arising from the variation of longitude of places having been considered objectionable', anticipates from the submarine telegraph to France an extension of this practice, and proposes the question, 'What meridian shall be determined on for universal adoption?'

Surely, Sir, we may rather, on the contrary, hope that the facilities thus afforded for its exaggeration will make the unreasonableness of the custom referred to more glaring, and will lead to a return to the only true and simple rule of keeping our clocks right instead of keeping them wrong, as we all now do, with the exception of a few wise men in the west. The absurdity of calling it noon at a quarter to 12 is not strikingly obvious; but every one must see the absurdity of calling it noon at sunrise in one place, at sunset in another, at midnight in a third.

It so happens that the world takes 24 hours to revolve on its axis. The fact may be 'considered objectionable'; but so long as it remains unaltered it is simply impossible that it should be the same hour in two different places at once; and 'uniformity of time' is as impracticable as uniformity of locality.

Truth and common sense cannot long be violated with impunity; and great inconvenience and confusion will inevitably arise (and shortly, too, with the aid of the submarine telegraph), unless we return to the plain fashion of calling it noon when it is noon, and not when it is not, and accept the variation in longitude in places as an irresistible fact, instead of voting it 'objectionable', and pretending to ignore it.

The convenience of central regulation need not, of course, be lost by keeping true local time. A timekeeper 10 minutes west of London, on receiving his electric intimation that it is noon at Greenwich, sets his clock at 10 minutes to 12, which is just as easy as setting it to 12. The only persons who derive any real benefit from the so-called 'uniformity' are the clerks who settle the railway time-tables, who are thereby saved some 20 minutes' trouble in calculating the allowance of time to be made as trains proceed east or west. This appears hardly a sufficient ground for bewildering all the timepieces and headpieces in Europe; but these gentlemen can do with us what they please, and we must be content with humbly entreating what we dare not demand, that they will cease from their desperate attempts to 'annihilate both Space and Time' (which have not even the laudable effect of 'making two lovers happy'), and allow the old legitimate King Time to resume his place in our clocks and bosoms.

Which only goes to prove that *The Times* has long been able to embrace many opinions, even when the dispute is over a subject embodied in its own name.

In January 1852 *The Illustrated London News* announced Airy's plans for time signals by electric telegraph, noting that the only towns of consequence still holding out against Greenwich time were mainly in the east and the west of the kingdom – Norwich, Yarmouth, Cambridge, Ipswich, Colchester, Harwich; and Oxford, Bristol, Bath, Portsmouth, Exeter, Dorchester, Launceston, Falmouth.[46] In Oxford, according to B. L. Vulliamy, clockmaker to the Queen, the inconvenience of two different times was met on the great clock on Tom Tower at Christ Church by employing two minute hands, one set to Oxford time, one to Greenwich.[47]

A place where the debate waxed strongly was Exeter. On 13 November 1851 H.S.E. of Exeter – almost certainly the publisher of the map – wrote to *The Times* to say that, at a recent meeting, the Town Council had resolved by 16 votes to 5 that 'public clocks in this city should show and strike Greenwich instead of local mean time'. But the Dean and Chapter had refused to alter the cathedral clock on which all other public clocks relied, one objection being that Plymouth had altered theirs and then altered back. (In fact, this was because the Plymouth tide tables were based on local time but, as was pointed out, this could very easily be changed in the future.) H.S.E. demanded legislation on the matter of legal time.[48]

In August 1852 the electric telegraph was about to reach Exeter – and, indeed, Plymouth also – and the local authorities decided to raise the subject of time once again. Sir Stafford Northcote (1818-87), a prominent local resident in favour of the change who rose to become Chancellor of the Exchequer in Disraeli's Government of 1874-80, wrote to the Astronomer Royal in August 1852 to ask his views on the whole question – and specifically to ask whether he was the Chronos who wrote the letter of 7 October 1851 quoted above.[49] No! said Airy, he was not Chronos. 'My opinion on the question is not very distinctly formed.' He then gave about three lines in favour of the change, followed by seven pages giving the opposite view, including the words: 'If I had the power of legislating in a comprehensive way, I should certainly preserve local time in every place'[50] – which today seems a curious attitude in view of his subsequent declaration that the dissemination of accurate time was one of the prime functions of the national observatory, and his boast only nine months later: 'I cannot but feel a satisfaction in thinking that the Royal Observatory is thus quietly contributing to the punctuality of business through a large portion of this busy country.'[51]

Immediately after this exchange, the *Western Luminary* of 31 August 1852 reported that another Council meeting on the subject had been held. Towns in the West Country which did keep Greenwich time – Taunton, Tiverton, Torquay, for example – felt no inconvenience; in other towns in the kingdom where Greenwich time had been adopted the change had been forgotten only a few days after the alteration had been made; for countrymen, the change from apparent (sundial) time to mean (clock) time had had an even greater impact than the change from Local Mean Time to Greenwich Mean Time.[52] On 28 October a public meeting was held at the Guildhall, Exeter, where it was unanimously resolved that public clocks should be altered. The Dean gave way: the cathedral clock was advanced 14 minutes to Greenwich time on 2 November 1852, the day after the first regular daily time-signal was sent down the railway telegraph lines direct from Greenwich.[53]

In Bristol, also, there was considerable debate, turning partly upon the business of tide tables. The *Bristol Times* of 20 March 1852 reported that no serious inconveniences had arisen at Newport, Swansea, Southampton or Liverpool. Bristol therefore made their decision to regulate clocks to Greenwich time more than a month before Exeter, at a meeting of the Council on 14 September 1852, though three inveterate admirers of ancient ways protested against the innovation. It is said that one of the more determined antiquarian councillors continued to regulate his affairs by local time for many years.[54]

In Plymouth a meeting of the Town Council took place on the subject the following day. Plymouth, unlike Bristol, adopted no resolution but some of the debate reported in the local paper had a surprisingly modern ring to it:

> Mr. W. MOORE disapproved of the movement, for workmen would thereby be enabled to leave work 16 minutes earlier every day, but he was sure employers would not be able to get them to come 16 minutes earlier in the morning *(a laugh)*.
>
> Mr. R. RUNDLE so far differed from the last speaker, that he believed tradesmen and others would be very glad of a change which would establish an uniform time throughout the Kingdom. He suggested that the Mayor should communicate with the Chief Magistrate of Devonport on the subject.
>
> Mr. W. F. COLLIER remarked it would be well if the Mayor could, at the same time, cause Railway trains to arrive at the hours at which they were due *(hear, hear)*.[55]

Time signals to private subscribers

From the early days of the Greenwich Time Service the telegraph companies were prepared to rent a private wire specifically for receiving the Greenwich time signal, a service particularly appreciated by public bodies and chronometer-makers in London and the provinces. At the time of the transfer of the telegraphs to the State (1870) the Post Office took over the contracts then in force,

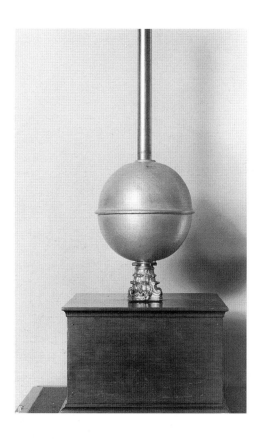

Above, left and right. Small time-ball, about 1855, and galvanometer, about 1900, examples of apparatus often set up in jewellers' windows so that the public could get the Greenwich time signal.

Opposite.
A hatter's time signal at George Carter & Sons Ltd., 211-17 Old Kent Road, London, S.E. From the *South-East London and Kentish Mercury*, 27 Feb.1975.

considerably adding to their number thereafter. The charges were first published in the *Post Office Guide* of 1 January 1873, varying according to the subscriber's distance from the main post office. In London within two miles of the GPO at St Martin's-le-Grand, for example, subscribers could receive an hourly signal for £15 a year. Provincial subscribers ¼ mile from their local Head Post Office paid £12 annually for the 10 a.m. current, £27 for the 1 p.m. current; if a mile away, the charges were £17 and £32 respectively. The GPO offered this service until 1927.

Subscribers had to provide their own terminal equipment – for example, time-balls on buildings, small time-balls in shop windows, bells, galvanometers, etc. One of the more bizarre time signals was that of the hatter George Carter of Old Kent Road, London S.E., which, from about 1900, consisted of an elephantine silk hat outside the shop which, after climbing slowly up a mast, fell at precisely 1 p.m. on receipt of the Greenwich time signal. In response to fashion, the top hat was replaced after the First World War by a figure with a bowler hat, which was raised and let fall at one o'clock daily.[56]

In 1876 the Standard Time Company was formed by the chronometer-makers Barraud & Lunds of Cornhill, London, principally to exploit a method of synchronizing clocks invented by J. A. Lund. The company received the hourly time signal from Greenwich by direct private wire and redistributed it to subscribers by its own overhead wires in the London area. Though the signal was principally designed to synchronize the subscribers' master clocks (which could then be used to drive slave clocks) by the forcible correction of the minute hands every hour, it could also be used to operate any form of time signal, audible or visual. Though the bombing of London in 1941 forced the company to abandon overhead wires, the system continued to operate until 1964.[57] Somewhat similar systems operated in Edinburgh, Glasgow and Liverpool, initially using the time signals from local observatories. These British systems were reported to be far superior to the pneumatic system started in Paris about the same time, where the clocks furthest from the depot were said to be appreciably slower than those which were closer, owing to the slow build-up of pressure in the tubes.[58]

Legal time

Although by 1855 98 per cent of the public clocks in Great Britain were set to GMT, there was still nothing on the statute book to define what was the time for legal purposes. Say someone died in Inveraray at 11.50 p.m., just *before* midnight Sunday Inverarary time, but *10* minutes *after* midnight Monday Greenwich time: did he die on Sunday or Monday? The answer to that question could have important legal consequences in insurance, inheritances, etc. The case of *Curtis v. March*, 25 November 1858, tested this matter:

> This was an action of ejectment which was entered for trial before Watson B [Baron of the Exchequer] at the last Dorchester Assizes. The time appointed for the sitting of the Court was 10 o'clock A.M., and the learned Judge took his seat on the bench punctually at 10 by the clock in Court. The cause was then called on and the plaintiff's counsel commenced his address to the jury, but as the defendant was not present and no one appeared for him, the learned Judge directed a verdict for the plaintiff. The defendant's counsel then entered the Court and claimed to have the cause tried, on the ground that it had been disposed of before 10 o'clock. At that time it wanted one minute and a half to 10 by the town clock. The clock in Court was regulated by Greenwich time, which was some minutes before the time at Dorchester. . . '[59]

On appeal, Chief Baron of the Exchequer Pollock, sitting with Barons Watson and Channell, reversed Watson's original decision:

> Ten o'clock is 10 o'clock according to the time of the place, and the town council cannot say that it is not, but that it is 10 o'clock by Greenwich time. Nor can the time be altered by a railway company whose railway passes through the place nor by any person who regulates the clock in the town-hall.[60]

The decision reached in 1858, specific to the time of sitting of courts, was held also to define legal time for other purposes in Great Britain until 1880: 'The time appointed for the sitting of a Court must be understood as the mean time at the place where the Court sits, and not Greenwich time, unless it be so expressed.'[61]

On 14 May 1880 *The Times* published a letter from 'Clerk to the Justices':

> During the recent elections many members of Parliament and the officials conducting elections must have been much troubled to decide what was the correct time to open and close the poll.
>
> Greenwich time is now kept almost throughout England, but it appears that Greenwich time is not legal time. For example, our polling booths were opened, say, at 8.13 and closed at 4.13 p.m.
>
> This point as to what is legal time often arises in our criminal courts, but has hitherto escaped a proper decision and discussion. Will not some new M.P. take up this point and endeavour to get an Act making Greenwich time legal time![62]

On 1 June of the same year the Statutes (Definition of Time) Bill was read for the first time in the House of Commons, sponsored by Dr. Cameron, Mr. David Jenkins, and Mr. Ervington. It was referred to a committee whose report was laid before the House by Mr. Playfair on 5 July. It passed through the remaining stages in both Houses almost without debate and received the Royal Assent on 2 August 1880.[63] It stated firmly, and at last:

> Whenever any expression of time occurs in any Acts of Parliament, deed, or other legal instrument, the time referred shall, unless it is otherwise specifically stated, be held in the case of Great Britain to be Greenwich mean time, and in the case of Ireland, Dublin mean time.

A story told about 1950 by Mr. J. H. Garner, a superintendent of the Central Telegraph Office, epitomizes the way in which the dissemination of a standard time system brought people together. After the Rugby time signal was instituted in 1927, the telegraph lines once used for time signals were 'recovered' so that they could be used for other purposes:

> Large numbers of lines were nominated as chronopher lines, with a number of reserves, and great care was always taken that all such nominated lines were tested and maintained so that they were all in good order for the time signal. One such reserve was the TS-BS London-Bristol – proved and checked every day – and regularly at 9.0 a.m. the time signal was transmitted, as was thought, to a centre on the north coast of Somerset. . .
>
> An enquiry was begun into whether the cost of maintenance of such lines was justified and in due course the Bristol line came under review, with the remarkable result that no trace of any circuit beyond Bristol could be found. Enquiries established that, years before, the line ran to a Customs and Excise and Coastguard centre on the Somerset coast, but changes had brought the centre into disuse and now only one coastguard remained.
>
> The engineers, being very thorough, traced and recovered what was left of the circuit and found its termination at the coastguard house where, boldly shown in the office window, was the well-known rod and ball with a notice that at 9.0 a.m. precisely the ball would fall to indicate Greenwich time.
>
> The coastguard admitted that he had not received a time signal for many years but the inhabitants expected a signal from him so he knocked the ball down with a stick each day. When asked how he got the time he said that by standing on a chair and looking along and across the road he could see by the 'Brown's' big clock when to knock the ball down.
>
> He was told to cease the practice and the engineers, being curious, went along to examine Brown's clock. They found it a well-built English clock, and only 15 seconds slow. Asked how he kept the clock at correct time, Mr Brown said: 'Oh, that's easy. By standing on my stool and peering sideways along and across the road, I can just see in the coastguard's front window the brass ball fall on the rod.'[64]

FIVE

A prime meridian

1790 - 1884

Opposite. The Prime
Meridian of the world,
Old Royal Observatory,
Greenwich, with the
Accurist Millennium
Countdown Clock.

Before the advent of railways and telegraphs, the only non-nautical requirement for Greenwich time known outside Britain was specifically geographic – in the precise measurement of the difference of longitude between observatories. Difference of longitude can be determined astronomically or geodesically (by trigonometrical-survey methods) or, as we have seen, by the transport of chronometers. One of the earliest examples of this last method took place during the geodetic operation to connect Paris and Greenwich Observatories in 1784-8, instigated by Cassini de Thury and conducted on the English side of the Channel by Major-General William Roy, FRS. In September 1785 Maskelyne sent his assistant Joseph Lindley by post-chaise and cross-channel packet to Paris and back carrying eight of John Arnold's chronometers, yielding a difference of longitude of 9 minutes 19.8 seconds, only about a second too small and agreeing well with the existing astronomical determinations and the geodetic result.[1] In 1825 a series of rockets was used by John Herschel and Col. Sabine to connect Paris and Greenwich.

In the 1830s, the German mathematician Karl Gauss set up an international association whereby observatories in Europe made observations of terrestrial magnetism on fixed 'term days', all at precisely the same moment – by Göttingen mean time.

The chronometer method was used for longitude determination of observatories until the coming of the electric telegraph. For example, in 1845 over sixty chronometers were sent sixteen times backwards and forwards between Altona near Hamburg and Pulkowa near today's St Petersburg, and the following year forty chronometers went the same number of times between Altona and Greenwich. Chronometers were sent across the Atlantic many times to determine the longitude difference between Harvard and Liverpool Observatories, from which the difference of longitude between Harvard and Greenwich was accurately determined. In 1844 the longitude of Valentia Island and the west coast of Ireland was found in this way, under Airy's superintendence.

The use of the electric telegraph for this purpose was first suggested by the American astronomer S. C. Walker and first used in the USA about 1844. As we have seen, the telegraphic connections between Greenwich and the Continent were suggested by Airy in 1851, connection with Brussels being established in 1853, and with Paris in 1854. The longitude of Valentia was redetermined by telegraph in 1862. Telegraph signals do take a finite time to travel along the lines, depending upon the distance and the number and type of relays. Airy reported a time of passage of $\frac{1}{12}$ second from Greenwich to Paris in 1854 and $\frac{1}{8}$ second (nearly) to Valentia in 1862.[2]

The Atlantic cable

The first successful submarine cable was laid across the English Channel in 1851. Wales and Scotland were linked with Ireland in 1852, England with Belgium and Denmark in 1855. By 1860 London was connected with the Indian subcontinent, one of the longest submarine cables being between Malta and Alexandria, 1,565 miles. But the really exciting prospect was a cable – perhaps more than one - between Europe and North America. Its main protagonist was the great American Cyrus W. Field (1819-92), whose untiring efforts provided the impetus throughout.

The Atlantic Telegraph Company was formed in October 1856 specifically to establish telegraphic communication between Newfoundland and Ireland. In 1857 half the cable to be used was embarked in the US steam frigate *Niagara* at Birkenhead, the other half in HMS *Agamemnon* at the works of the Telegraph Construction and Maintenance Company in East Greenwich, within sight of the Royal Observatory. After an unsuccessful attempt in 1857, a second was made in the spring of 1858. On 5 August the first telegraph message was sent across the Atlantic, causing great jubilation on both sides of the ocean. However, by 3 September, less than a month later, communication failed and could not be re-established.

The American Civil War between 1861 and 1865 brought only a temporary halt in Field's efforts to achieve a transatlantic cable. More capital was raised and heavier cable manufactured at Greenwich. For two ships to lay the cable had not proved entirely satisfactory, so it was decided to use the only ship then capable of embarking all the cable needed – the *Great Eastern*, 22,500 tons, the conception of Isambard Kingdom Brunel, built at Millwall on the Thames and the world's largest ship. Because of her draught (nearly 35 feet when

Isambard Kingdom Brunel, photographed in front of the launching chains of the SS *Great Eastern* at Millwall, on the Thames, in 1857.

loaded) the *Great Eastern* had to lie at Sheerness to embark the cable, which had been sent down-river by lighter from Greenwich.

On 23 July 1865 she left Valentia, Ireland, paying out her cable, escorted by HM Ships *Terrible* and *Sphinx* and having on board Professor William Thomson, later Lord Kelvin (1824-1907), one of the greatest names in Victorian science. Alas, on 2 August, after 1,025 nautical miles of cable had been paid out, it parted and could not be recovered. The next year, however, a new cable was successfully laid by the *Great Eastern*, taking fourteen days from Valentia to Heart's Content in Newfoundland. To complete the triumph, the *Great Eastern* successfully grappled the 1865 cable, spliced it on to cable remaining on board, and thus provided a second cable link across the Atlantic.

One of the factors leading to this success was that during the 1866 lay, at the suggestion of Captain Anderson, the Greenwich time signal was received by the *Great Eastern* twice daily by telegraph via London, Holyhead, Dublin, Valentia,

Top. The *Great Eastern*, then the largest ship ever built, at 18,915 tons, shown during her Atlantic cable-laying voyage, 1866, painted by Henry Clifford. Below. Splicing the Atlantic cable aboard the *Great Eastern*, from W. H. Russell, *The Atlantic Cable* (1875).

and the cable she was laying, thus enabling her to find her longitude exactly.[3] This seems to be the earliest example of a ship at sea receiving a time signal by other than visual means. One of the earliest uses of the new cable was to redetermine the longitude difference between the observatories of Greenwich and Harvard University at Cambridge, Mass. This was conducted in October 1866 by Dr B. A. Gould of the US Coast Survey, in co-operation with Airy.[4]

European railways

On the continent of Europe, the railways brought the same problems of timekeeping that they had in Britain. In general, though local mean time was kept by the passengers, the trains in each country were run according to some central time. In France, for instance, clocks inside railway stations were kept to *l'heure de la gare* which was 5 minutes slow on Paris time, while clocks outside the station were kept to local time, *l'heure de la ville*. Belgian trains ran to Brussels time, Dutch trains to Amsterdam time. In Germany railway officials kept any one of five times – those of Berlin, Munich, Stuttgart, Karlsruhe or Ludwigshafen. Passengers, however, kept strictly to local time and there were posts set alongside the rails marking each minute's change of time. It was apparently customary for watches to be altered in ten-minute steps during the journey. This fixation on local time, encouraged by the German astronomers, was only abandoned, and Berlin time adopted throughout, when Count von Moltke pointed out the military consequences of this lack of standardization.[5] All of which must have been very confusing to long-distance travellers.

US railroads

If the need for co-ordination of timekeeping was evident in Great Britain, with a maximum longitude difference equivalent to 30 minutes of time, how much greater were the potential difficulties in the United States, where the difference between east and west amounted to more than 3½ hours? However, it was not this difference that drew attention to the problem, but the fact that the many railway companies which sprang up after the end of the Civil War each kept its own time, as did each town and city on the way, to the great inconvenience of the travelling public. For instance, a traveller from Portland, Maine, on reaching Buffalo, NY, would find four different kinds of 'time': the New York Central railroad clock might indicate 12.00 (New York time), the Lake Shore and Michigan Southern clocks in the same room 11.25 (Columbus time), the Buffalo city clocks 11.40, and his own watch 12.15 (Portland time).

At Pittsburgh, Penn., there were six different time standards for the arrival and departure of trains. A traveller from Eastport, Maine, going to San Francisco, was obliged, if he was anxious to have correct railroad time, to change his watch some twenty times during the journey.[6]

The first observatory in the USA to distribute time seems to have been the Naval Observatory in Washington, from August 1865.[7] Then in 1869 Professor S. P. Langley of the Allegheny Observatory near Pittsburgh instituted a distribution system which covered a very considerable area, from Philadelphia and New York in the east, Lake Erie in the north and Chicago in the west. But three separate time signals were necessary – at Pittsburgh time for the local watchmakers and jewellers; at Altoona time (10 minutes fast on Pittsburgh) for the Pennsylvania Central Railroad to Philadelphia and eastwards; and at Columbus time (13 minutes slow on Pittsburgh) for the Pittsburgh, Fort Wayne and Chicago railroad.[8] In his description of the system, Langley put in the strongest plea for a uniform standard of time, noting that the managers of the railroads connecting New York, Philadelphia, Pittsburgh and Chicago were then (1872) contemplating the use of one standard time, that of the meridian of Pittsburgh, for all trains between the cities concerned.

Meanwhile, there were other thoughts about the matter of standardizing time for railroad and other purposes in the United States. In 1870 a 107-page pamphlet was published entitled *A System of National Time for Railroads*, written by Professor Charles Ferdinand Dowd (1825-1904), Principal of Temple Grove Ladies' Seminary in Saratoga Springs, NY. Dowd's pamphlet was a result of discussions at the Convention of Railroad Trunk Lines in New York City in October 1869. To obviate the inconvenience of having some eighty different time standards on the various US railroads – not to mention the fact that each station had to maintain its own local time as well – Dowd proposed a scheme identical in principle with the standard time system used all over the world today: that, for Railroad Time purposes in the USA, there should be four standard meridians 15° (= 1 hour) apart, the eastern one being the Washington meridian. These meridians would be the centres of four time zones; in each zone the time adopted would be uniform, and it would change by one hour when passing from one zone to the next. The boundaries of each zone, though approximating to the appropriate meridians, were to be adjusted to take into account local state or county boundaries and the areas served by individual railroads. This meant that, in theory, the minute and second hands of all clocks would show the same time, only the hour hands being different. But Dowd's first scheme differed from the zone-time sys-

Charles F. Dowd, of Saratoga Springs, NY, who first devised the zone-time system used all over the world today. From *Harper's Weekly*, 29 Dec. 1883.

tem of today in one particular – it was based on the meridian of Washington, specifically that passing through the transit instrument at the US Naval Observatory. (It is of interest that, in the Act of 28 September 1850 making the appropriation for the US Nautical Almanac, Congress made the proviso 'that hereafter the meridian of the observatory at Washington shall be adopted and used as the American meridian for all astronomical purposes, and that of the meridian of Greenwich shall be adopted for all nautical purposes'.)

The convention agreed wholeheartedly in principle with Dowd's proposals and he was asked to go and work out the scheme in detail. For instance, it was obviously impracticable for the limits of each zone to conform *exactly* to the appropriate meridian or one could have different times being kept in different rooms of the same house. To continue the story in Dowd's own words:

I at first took the national meridian of Washington, and having divided the country into three 15-degree belts, patiently marked out the longitude of some 8,000 stations along some 500 railroad lines, and had a map engraved showing the hour sections and the proposed Standard versus the actual time at each station. With this map was incorporated and published a pamphlet of 100 octavo pages, which I sent to all railroad men and others in this country who would likely feel an interest in the work.

In 1871 I presented the subject to the North East Railroad Association in Boston, and to others elsewhere. Then came suggestions that brought to my mind that it would be better to adopt the nautical meridian of Greenwich, and by using its fifth hour or the 75th meridian west longitude as the Prime American standard, I laid out the hour sections upon that basis. In the spring of 1872 I went to St. Louis and attended a meeting of the Western Railway Association, and to Atlanta, Georgia, to confer with superintendents of Southern railways.[9]

Apparently Dowd learned from these meetings that using the Washington meridian projected the other meridians too far west to suit the needs of the Eastern and Central sections, which felt their convenience and that of 'the borderline cities' should also be considered. So, in May 1872, Dowd changed the system and moved all the hour sections approximately two degrees east so that they now conformed with integral numbers of hours west of the Greenwich meridian. He summarized his proposals as follows:

Explanation of standards

The time of the 75th Meridian [west of Greenwich] is adopted as the standard time for all roads east of Ohio and the Allegheny Mountains;

and the time of the 90th Meridian for western roads situated anywhere in the Mississippi valley. These times may be designated Eastern and Western times, their difference being just one hour. Following westward still, the next hour standard falls in the Rocky Mountain District, and hence is of little avail. But the third hour standard, or the time of the 120th Meridian, is very central and convenient for roads on the Pacific coast. Again the fifth hour eastward is adopted as the standard time of England, and is the basis of longitude on all marine charts.

Saratoga Springs, NY,   C. F. Dowd.[10] May 15, 1872.

As might be expected, Dowd's proposals gave rise to much debate, and many alternative suggestions were put forward, in particular those of 1879-80 where the use of a single standard time, that of the 90th meridian west of Greenwich (which almost passes through St Louis and New Orleans), was seriously considered. The full story is too long to give here and has in any case been lucidly summarized elsewhere.[11]

Eleven years and many railroad conventions later, Dowd's 1872 plan was adopted virtually unchanged by the railways of the USA and Canada. At noon or before on Sunday 18 November 1883 public clocks all over North America were altered to the 'new standard of time agreed upon, first by the railroads, for the sake of the uniformity of their schedules, but since generally adopted by the community through the action of various official and corporate bodies as an obvious convenience in all social and business matters'.[12]

Newspapers all over the country reported reactions. The *New York Herald*, for example, writing on change-over day, pointed out that someone going to church in New York that day would discover that the noon service had been curtailed by almost four minutes, while every old maid on Beacon Hill in Boston would rejoice that night to discover she was younger by almost sixteen minutes. On the other hand, everyone in Washington would be eight minutes *older*. In the central zone, the new time in Chicago was 9½ minutes slower than the old; in Cleveland, Ohio, not far from the border of the zone, 32 minutes slower.

The next day, in an article headlined 'Horometric Harmony', the *Herald* reported that great crowds had watched the dropping of the time-ball on the top of the Western Union building. At the Church of Our Father in Brooklyn a sermon was preached – duly reported in about six column-inches – on 'Changing and Keeping Time', with the text from Joshua 10:13: 'And the sun stood still, and the moon stayed, until the people had avenged themselves upon their enemies. Is not this written in the book of Jashar? So the sun stood still in the midst of heaven, and hasted not to go down about a whole day.'[13]

Harper's Weekly, besides reproducing a portrait of Dowd and a map illustrating the new time zones, outlined the advantages in the flowery prose of the 1880s:

> On Saturday, the 17th of November, when the sun reached the meridian of the eastern border of Maine, clocks began their jangle for the hour of twelve, and this was kept up in a drift across the continent for four hours, like incoherent cowbells in a wild wood.
>
> But on Monday, the 19th – supposing all to have changed to the new system on the 18th – no clock struck for this hour till the sun reached the seventy-fifth meridian. Then all the clocks on the continent struck together, those in the Eastern Section striking twelve, those in the Central striking eleven, those in the Mountains striking ten, and those in the Pacific striking nine.
>
> The minute-hands of all were in harmony with each other, and with those of all travellers' watches. Time-balls everywhere became perfectly intelligible, and the bliss of ignorance was no longer at a premium.[14]

So, in 1883, the United States and Canada adopted a time standard based upon the Greenwich meridian. Although it was not until 1918 that an Act of Congress legalized standard time all over the United States, the civil population nevertheless adopted 'Railroad Time' almost spontaneously, as had happened in Britain thirty years before: 85 per cent of US towns of over ten thousand inhabitants had done so by October 1884. There were exceptions. Detroit, Michigan, for example, on the borderline between the Eastern and Central zones, continued to keep local time until 1900 when the

US time zones, in 1883 and today. After C. J. Corliss, *The Day of Two Noons* (Washington, DC, [1941]).

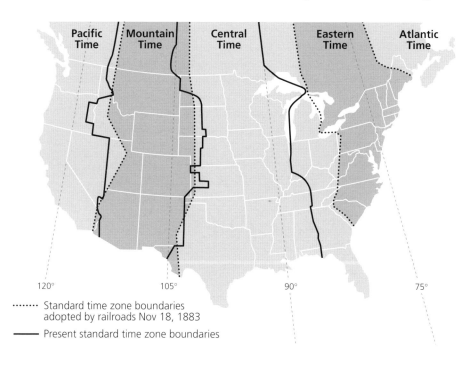

.... Standard time zone boundaries
adopted by railroads Nov 18, 1883

———— Present standard time zone boundaries

City Council decreed that clocks should be put back twenty-eight minutes to Central Standard Time. Half the city obeyed, half refused. After considerable debate, the decision was rescinded and the city reverted to Sun time. A derisive offer to erect a sundial in front of the city hall was referred to the Committee on Sewers.[15] Then, in 1905, Central time was adopted by city vote. In May 1915 this was changed by ordinance to Eastern Standard, a decision upheld by popular vote in August 1916.[16]

Though, as we shall see, many others – particularly the Canadian, Sandford Fleming – should be given the credit for encouraging the adoption of Standard Time on a worldwide basis, it is to Charles F. Dowd that we are indebted for the principle. Ironically, he met his death beneath the wheels of a railroad locomotive in Saratoga, New York, in 1904.

A prime meridian

Meanwhile, particularly since 1870, geographers and scientists of allied disciplines from all nations had been giving their attention to the possibility of fixing a common zero for longitude and time-reckoning throughout the globe. And it was the former – the prime meridian – which was the first to be discussed. As we have seen, the first astronomer to determine differences of longitude seems to have been Hipparchos. For his first meridian – or prime meridian, or *méridien-origine*, or *méridien initial*, or *meridiano iniziale*, or *Nullmeridien* – he used Rhodes, where he was observing. But Ptolemy, following Marinus of Tyre, adopted a meridian through the *Insulae Fortunatae*, the Canary Islands, which seemed to mark the western boundary of the world, whereas, to the east, there seemed to be no such boundary.

With the voyages of discovery in the fifteenth and sixteenth centuries, a new interest arose. In 1493 Pope Alexander VI laid down a line of demarcation between the spheres of influence of Portugal and Spain 100 leagues west of the Azores and the Cape Verde Islands; after protests by Portugal, this was shifted, in the Treaty of Tordesillas (1494), to a meridian 370 leagues west of the Cape Verde Islands which was itself used on some maps and charts as a prime meridian. On 13 July 1573 Philip II of Spain issued two Ordinances concerning the measurement of longitude throughout the Spanish empire. The first (No. 62) stipulated that all longitudes should be computed from the meridian of the city of Toledo and, contrary to the practice of the ancients who measured east from the Canaries, Spanish longitudes should be measured westwards 'because to proceed in this manner is more natural and conforms to the discovery of the Indies which God was pleased to grant us'.[17]

Ordinance No. 67 ordered colonial governors to take every opportunity to have eclipses and other phenomena capable of yielding difference of longitude observed, the results to be reported to the Chief Cosmographer of the *Casa de Contratación* (House of Trade) at Seville, so that the longitudes could be computed to complete the *Padrón Real*, the inventory of data about the New World which enabled the plotting of the official Spanish nautical charts.

Later, cartographers such as Mercator and Ortelius began to choose various islands in the western ocean – in the Canaries, Madeira, the Cape Verdes, and the Azores – for reasons described by William Blaeu in the Latin cartouche of his globe of 1622:

> . . . But in our days a good many think this starting point ought to be based on nature itself, and have taken the direction of the magnetic needle as their guide and placed the prime meridian where that points due north. But that these are under a delusion is proved by that additional property of the magnetic needle through which it is no standard for the meridian, for itself it varies along the same meridian according as it is near one land mass or another. . . .

Blaeu goes on to explain that, for his own globe,

> . . . following in the steps of Ptolemy, [we] have chosen the same islands and in them Juno, commonly called *Tenerife*, whose lofty and steep summit covered with perpetual cloud, called by the natives *El Pico*, shall mark the prime meridian. In that way we have differed barely a quarter of a degree from the longitude of the Arabs who chose the extreme western shore of Africa [C. Verde], and I thought it well to point this out.[18]

In April 1634 Cardinal Richelieu called a conference of eminent European mathematicians and astronomers in Paris to consider the question of a prime meridian to be recognized by all nations. The choice fell upon Ptolemy's Fortunate Isles, more closely defined as the west coast of the island of Ferro (or Ile de Fer, or Hierro), the westernmost of the Canary Isles. However, the Thirty Years War was in progress and, judging by the terms of Louis XIII's decree of 1 July 1634, the motive behind the calling of the conference seems to have been at least as much political as scientific:

> French ships are not to attack Spanish or Portuguese ships in waters lying east of the First Meridian and north of the Tropic of Cancer. In order that this first meridian may be more clearly known than it has for some time been, the Admiral of France has consulted persons of knowledge and experience in navigation. The King in consequence forbids all pilots, hydrographers, designers or engravers of maps or terrestrial globes to innovate or vary from the ancient meridian passing through the most westerly of the *Canary Islands*, without regard to the novel ideas of those who recently fixed it in the Azores on the supposition that there the compass does not vary, for it is certain that this happens also in other places that have never been taken for the meridian.[19]

In 1724 Louis Feuillée was sent by the Paris Academy to make actual measurements to determine the longitude of Paris based on Richelieu's prime meridian. Depending upon Feuillée's astronomical observations taken on the neighbouring island of Tenerife, the figures eventually published in 1742 were: Paris (Notre Dame) 20° 02′.5 east of the western point of Ferro; London (St. Paul's) 17°37′.5 E.[20] However, there was still no general agreement on a prime meridian and each nation tended to please itself, generally using its capital city or principal observatory. Navigators in their logs generally used the point of departure on any specific leg of the voyage, 15° 27′ west of the Lizard, or 26° 32′ east of the Cape of Good Hope, for example. Most sea charts in the eighteenth century had only a single longitude scale (if any at all) whose origin depended upon the national origin of the chart. The French, however, logical as always, put multiple longitude scales on many of the charts in the great series of official sea-atlases, *Neptune François,* so that the navigator could at will plot his position relative to Tenerife, Ferro, the Lizard and Paris. Incidentally, the same charts also had multiple scales of distance – of Breton leagues, French leagues, English leagues, as well as nautical miles – symptomatic of the confusion that prevailed before the international standardization of weights and measures.

We have seen that the publication of the British *Nautical Almanac* in 1767 meant that Greenwich began to be used as the prime meridian on maps and charts of many nations from the late 1700s. The first series of charts to use the Greenwich meridian systematically was J. F. W. DesBarres's *Atlantic Neptune* covering the east coast of North America from Labrador to the Gulf of Mexico, first published in 1784. DesBarres's charts continued to be used as the primary source for most American charts for the next fifty years and, with the British *Nautical Almanac* itself, were probably the primary reason for the decision by the US Government to retain the Greenwich meridian for nautical purposes in 1850. In 1853 the High Admiral of the Russian Fleet cancelled the use of the nautical almanac specially prepared for Russia and, in its place, introduced to the Russian Navy the British *Nautical Almanac,* based on the Greenwich meridian, from which the *Morski Miesiatseslob* (Naval Almanac) was produced.[21]

International discussions

The first International Geographical Congress (IGC) took place at Antwerp in August 1871. One of the resolutions passed expressed the view that, for passage charts of all nations (but not necessarily coastal or harbour charts), the Greenwich meridian should be adopted as the common zero for longitude, and that this should

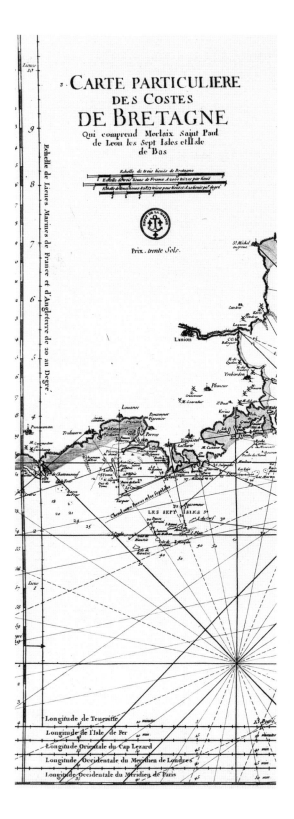

become obligatory within fifteen years. It was also recommended that, whenever ships exchanged longitudes at sea, they should be based on Greenwich. For land maps and coastal charts, however, each state should keep its own prime meridian.[22] In the discussion, M. Levasseur, one of the French representatives, generously said that, had it been the seventeenth or eighteenth century, the choice must have fallen on Paris, but that circumstances had changed: the majority of charts used at sea were now of British provenance and, furthermore, *le livre habituel du marin* was the British *Nautical Almanac*. (A few years later, it was said that the annual sales of the *Nautical Almanac* were 20,000 copies; whereas those of the *Connaissance des Temps* were 3,000 copies.)[23] M. Levasseur therefore supported the resolution in so far as sea charts were concerned.

The 2nd IGC in Rome in 1875 discussed the whole matter again without coming to any further conclusions. France, however, expressed a sentiment which was to be renewed time and again in subsequent discussions – that if England were to accept the metric system, then it would perhaps be courteous of France to accept the Greenwich meridian.[24] Table I shows that, only a dozen years later, the 1871 resolution had begun to take effect: twelve nations were counting their longitudes from Greenwich on newly published sea charts. On land maps, however, each nation continued to go its own way.

In 1876 there came a development from across the Atlantic, when a memoir called *Terrestrial Time* was published in Canada by Sandford Fleming (1827-1915), engineer-in-chief of the Canadian Pacific Railway. Born in Scotland, where he studied engineering and surveying, Fleming went to Canada in 1845. He started his career as a railway builder, becoming chief engineer of the Ontario, Simcoe and Huron (later Northern) Railway in 1852. In 1862 he took a similar post with the Intercolonial Railway and was appointed engineer-in-chief of the Canadian Pacific Railway, where he surveyed the Yellowhead Pass route (now followed by the Canadian National Railway) and was the first to demonstrate the practicability of the route through the Kicking Horse, Eagle, and Rogers Passes. Originally Fleming was interested only in the use of a 24-hour clock system, but then became fired with enthusiasm for the idea of a uniform time for the whole world – Terrestrial Time, or, as he later suggested, Cosmopolitan (and later still Cosmic) Time. He suggested an hour-zone system similar to Dowd's for use as local time for domestic purposes (he did not acknowledge Dowd's idea, though, being in the railroad business, he must have known about it), while his Terrestrial Time would be used by railroads, telegraphs, the sciences, etc.[25] Later, Fleming would be less enthusiastic about Dowd's standard-time system.[26]

In 1878-9 Fleming read two papers before the Canadian Institute, Toronto, entitled respectively 'Time-reckoning' and 'Longitude and time-reckoning'. The first was a rewritten version of his 1876 paper; the second, subtitled 'A few words on the selection of a prime meridian to be common to all nations, in connection with time-reckoning', put a strong case for the prime meridian being 180° from Greenwich, coinciding with the basic meridian used for today's International Date Line.[27] Fleming's two papers were considered so important that in June 1879 the British Government forwarded copies to eighteen foreign countries and to various scientific bodies in England.

On the whole, reaction was favourable. However, there were some notable exceptions, particularly among astronomers. Sir George Airy, writing in 1879, two years before his retirement, said that, first, he 'set not the slightest value on the remarks extending through the early parts of Mr Fleming's paper' [about the adoption of hour zones for local time and the establishment of 'Terrestrial Time']; and secondly, 'As to the need of a Prime Meridian, no practical man ever wants such a thing. If a Prime Meridian were to be adopted, it must be that of Greenwich, for the navigation of almost the whole world depends on calculations founded on that of Greenwich. . . But I as Superintendent of the Greenwich Observatory, entirely repudiate the idea of founding any claim on this: Let Greenwich do her best to maintain her high position in administering to the longitude of the world, and Nautical Almanacs do their best, and we will unite our efforts without special claim to the fictitious honour of a Prime Meridian.'[28] At Airy's suggestion, the Governor-General was told on 15 October 1879 that the British

Opposite. Multiple longitude scales on the charts in the official French atlases, *Le Neptune françois*. Detail from 'Carte particulière des Costes de Bretagne', c.1773, from an original of 1693.

Table I Prime meridians in use in the early 1880s on newly published maps and charts

Country	Prime meridian Sea charts	Land maps
Austria	Greenwich	Ferro
Bavaria	—	Munich
Belgium	Greenwich	Brussels
Brazil	Greenwich and Rio de Janeiro	Rio de Janeiro
Denmark	Greenwich, Copenhagen and Paris	Copenhagen
France and Algeria	Paris	Paris
Germany	Greenwich and Ferro	Ferro
Holland	Greenwich	Amsterdam
India	—	Greenwich
Italy	Greenwich	Rome
Japan	Greenwich	Greenwich
Norway	Greenwich and Christiania	Ferro and Christiania
Portugal	Lisbon	Lisbon
Russia	Greenwich, Pulkowa and Ferro	Ferro, Pulkowa, Warsaw and Paris
Spain	Cadiz (S. Fernando)	Madrid
Sweden	Greenwich, Stockholm and Paris	Ferro and Stockholm
Switzerland	—	Paris
UK and colonies	Greenwich	Greenwich
USA	Greenwich	Greenwich and Washington

Sources :
Sea charts: Borsari, F., *Il meridiano iniziale e l'ora universale* (Napoli, 1883), 60.
Land maps: Wheeler, G.M., *Report on the Third International Geographical Congress . . . Venice . . . 1881* (Washington, 1885), 30.

Government would not interfere in a matter which concerned social usages.[29] It seems strange that a man with such vision – to whom, more than anyone else, the country was indebted for the dissemination of Greenwich time – should hold such views at that date: only a year later, GMT was to become legal time in Great Britain; only three years later, a time-zone system was to be introduced throughout Canada and the USA; only five years later, it was agreed internationally that a prime meridian *was* needed, and that it should be Greenwich.

But among astronomers Airy was not alone. Professor Piazzi Smyth, Astronomer Royal for Scotland, while praising Fleming for his good intentions, condemned him for want of practicality. If there *had* to be a common prime meridian, why not the Great Pyramid in Egypt?[30] Professor Simon Newcomb, Superintendent of the American nautical almanac, was even more scathing. Asked in 1882 whether it was advisable for the USA to adopt a time system which would commend itself to other nations and be adopted by them ultimately, he answered, 'No! We don't care for other nations; we can't help them, and they can't help us.' To a second question as to whether the scheme for regulating time seemed to possess any features to commend itself to him, he said: 'A capital plan for use during the millennium. Too perfect for the present state of humanity. See no more reason for considering Europe in the matter than for considering the inhabitants of the planet Mars.'[31]

After which, it is almost a relief to note the reactions of the astronomers of Bologna who suggested Jerusalem as the prime meridian; and of Professor de Beaumont, a geographer from Geneva, who suggested a meridian straight through the Bering Straits, chosen so as to be an even number of degrees from both Paris and Ferro, with an anti-meridian (180° away) which would pass close to Rome, Venice and Copenhagen.[32] About this time also, the desirability of decimalizing the right angle and the hour (as had been originally conceived in the metric system) began to be discussed once again. All these views were in front of the delegates to the Third International Geographical Congress, which met in Venice in September 1881 and for which the establishment of a universal prime meridian and a uniform standard of time was high on the agenda. There was considerable discussion on these matters but the importance of this congress lay mainly in what stemmed from it – two special conferences about to be described.[33]

Sandford Fleming, engineer-in-chief of the Canadian Pacific Railway, who played a prominent part in forcing the adoption of standard time, and represented the Dominion of Canada at the International Meridian Conference in Washington in 1884.

The first of these was the Seventh International Geodesic Conference which assembled in Rome in October 1883. This conference of men of science is particularly important in that its conclusions formed the basis of the subsequent Washington conference which was primarily diplomatic in character. The Rome conference was attended by astronomers, geodesists and mathematicians. Great Britain and the USA were specially invited to send representatives in view of the main subject matter – longitude and time. Britain was represented by W. H. M. Christie (1845-1922), who had succeeded Airy as Astronomer Royal two years earlier.

They began their discussions with admirable dispatch, and the scientific objectivity, practical approach and lack of national prejudice are noticeable. 'The time has passed when pure science thought it beneath its dignity to concern itself when necessary with matters of general practical usefulness, and when governments and administrations thought they could manage without consulting men of learning on questions relating to science.' So ran the official report, which continued by noting that, though the conference was not empowered to make final decisions or definitive international agreements, nevertheless by its constitution it was likely to have an important influence on future decisions. And so it proved.

Resolution I epitomized the objective attitude of the delegates, stating that the unification of longitude and time was desirable as much in the interest of science as of navigation, commerce and international communications, and suggesting that the importance

The Paris Observatory, through which Cassini's meridian of 1672 passes.

of such a measure far outweighed any sacrifices which might have to be made.

Resolution II recommended the extension of the decimal division of the right angle to geodesic calculations, while retaining the sexagesimal system (the traditional degrees, minutes and seconds) for astronomy, navigation, maps and so on.

Resolution III was the fundamental one, concerning the actual selection of the prime meridian. In the absence of a natural zero for longitude (as the equator is the natural zero for latitude), the need for a prime meridian to be chosen arbitrarily was stressed and the scientific conditions to be fulfilled by such a meridian were discussed at length. Ideas of basing the new prime meridian on Ferro, on 20° west of Paris , on the Bering Straits, or of finding a 'non-natural' (neutral) meridian, were dismissed. The initial meridian must be defined by an observatory of the first order and, after noting that 90 per cent of navigators engaged in foreign trade already calculated their longitudes from Greenwich, the Conference proposed that Governments should adopt this as the initial meridian. Resolution IV, that longitudes should be counted in a single direction (from 0° to 360°), slipped through, but would be overturned the following year in Washington.

The conference then went on to discuss the unification of time, recognizing in Resolution V the usefulness of adopting a universal time for certain scientific needs and for internal use in railways, shipping lines, telegraphs and posts, to be used alongside local or national time which must necessarily continue to be used in civil life. In Resolution VI the conference recommended that the Universal Day should start at Greenwich noon, so that the new Universal Day and the Astronomical Day could be reconciled. This recommendation was also reversed at Washington.

In winding up, the conference hoped that, if the entire world was prepared to accept Greenwich as the prime meridian, Great Britain on her part might be prepared to conform to the metric system. It recommended a very early international convention devoted to the unification of longitude and time 'such as the United States government has proposed'.[34]

The International Meridian Conference, Washington, October 1884

As a result of the Venice geographical conference in September 1881, the United States passed an Act of Congress on 3 August 1882 authorizing the President to call an international conference to fix on a common prime meridian for time and longitude throughout the world.[35] On 23 October 1882 the State Department sent a circular letter to their representatives abroad asking whether such a conference would be welcome. The circular explained that, as the USA possessed 'the greatest longitudinal extension of any country traversed by railway and telegraph lines', it was very appropriate that she should call the conference. The response to the United States proposals for a conference in Washington was very favourable and the need for it had been underlined by the Rome conference of October 1883. On 1 December, 1883 therefore, invitations were sent to all nations in diplomatic relations with the US to send delegates – not exceeding three – to a conference to assemble in Washington on 1 October 1884. In view of future decisions, it is worth remarking that the North American railroads had adopted a standard time system based on the Greenwich meridian only eighteen days before the invitations were sent out.

On 1 October 1884 forty-one delegates from twenty-five countries assembled in Washington for the International Meridian Conference. (These figures do not include the delegate from Denmark who never arrived.) The majority were professional diplomats, though some countries sent scientific and technical representatives also. The conference was opened by the Secretary of State, the Hon. F. T. Frelinghuysen, in the name of President Arthur. The first business was to elect officers, Admiral Rodgers, USN, being elected President of the Conference, General Strachey of Great Britain, Prof. Janssen of France and Dr Cruls of Brazil being elected as secretaries.

The discussion on the main points then began, the aims of the conference having been defined in the Act of Congress: . . . for the purpose of discussing, and, if possible, fixing upon a meridian proper to be employed as a common zero of longitude and standard of time-reckoning throughout the whole world. . .' Over a period of a month the conference met on eight occasions and the report of their proceedings fills over two hundred pages. In the following summary the subtitles quoted give the wording of the resolutions as they were finally agreed, while the amendments and resolutions that were considered but not adopted are mentioned in the text if their importance warrants it. The voting country by country is analysed in Table II (p. 143).

I

'That it is the opinion of this Congress that it is desirable to adopt a single prime meridian for all nations, in place of the multiplicity of initial meridians which now exist.'

As a result of a suggestion from the Spanish delegate that discussion should be based upon the resolutions of the conference in

Rome, the conference proper opened with a resolution similar to Resolution II below – that Greenwich should mark the prime meridian – proposed by a US delegate. However, M. Lefaivre, a French diplomat, and his astronomical colleague, Prof. Janssen, suggested that this was prejudging the issue. As a result Cdr. Sampson of the USA proposed Resolution I which, after very little discussion, was adopted unanimously.

II

'That the Conference proposes to the Governments here represented the adoption of the meridian passing through the centre of the transit instrument at the Observatory of Greenwich as the initial meridian for longitude.'

The USA then once again proposed Resolution II. Prof. Janssen said that France was of the opinion that the purpose of the conference was to examine the principles upon which a prime meridian should be chosen, leaving the actual choice to a more technical conference. This provoked considerable discussion, the general view being that the whole purpose of the conference was not just to establish principles but actually to fix a prime meridian – as stated in the Act of Congress which initiated the conference. M. Lefaivre thereupon proposed another resolution – that the initial meridian should have 'a character of absolute neutrality. It should be chosen exclusively so as to secure to science and to international commerce all possible advantages and in particular especially should cut no great continent – neither Europe nor America.'

Discussion on this last resolution was long and went on for the whole of the session and well into the next. How could a meridian be absolutely neutral, argued Great Britain and the United States? It was very important that the prime meridian should pass through an astronomical observatory of the first order. Modern science, said Captain Evans (Great Britain), demanded such precision that one must abandon all ideas of establishing the meridian on an island (the Azores and Ferro had been suggested), on the summit of a mountain (say Tenerife), in a strait (say the Bering Strait), or as indicated by a monumental building (the Great Pyramid and the Temple of Jerusalem had both been put forward previously). This really left only the observatories of Paris, Berlin, Greenwich and Washington as satisfying scientific needs. Sampson (USA) added that it was important also 'to so fix and define it that natural changes of time may not render it in the least degree uncertain'. To define it as a certain number of degrees east or west of an established observatory did not thereby make it a neutral meridian, but merely disguised it. Surely one must also consider the practical point of view: with a new neutral meridian, *everyone* would have to change.

There was much discussion also on the metric system which the French suggested was a truly neutral system, '. . . and we are still awaiting the honour of seeing the adoption of the metrical system for common use in England'. No, said Cleveland Abbe, director of the Cincinnati Observatory, though the USA and England used the metric system as the standard in all important *scientific* work, it was nevertheless *not* entirely neutral, having been fixed by French measurement. 'Had the English, or the Germans, or the Americans taken the ten-millionth part of the quadrant of the meridian, they would have arrived at a slightly different measure. . . It was intended to be a neutral system, but it is a French system.'

And so it went on, the same arguments being put forward again and again, the earlier discussions being exclusively conducted by France on the one hand and the USA and Great Britain on the other – the one saying that the prime meridian must be 'neutral', the others saying that it could not be.

Then Sandford Fleming, the British delegate representing Canada, gave an address. He summed up the British view: 'A neutral meridian is excellent in theory but I fear it is entirely beyond the domain of practicability.' He then went on to read out the following table, showing the number and total tonnage of vessels using the several meridians listed for finding longitude:

Initial meridians	Ships of all kinds		Per cent	
	Number	*Tonnage*	*Ships*	*Tonnage*
Greenwich	37,663	14,600,972	65	72
Paris	5,914	1,735,083	10	8
Cadiz	2,468	666,602	5	3
Naples	2,263	715,448	4	4
Christiania	2,128	695,988	4	3
Ferro	1,497	567,682	2	3
Pulkowa	987	298,641	1½	1½
Stockholm	717	154,180	1½	1
Lisbon	491	164,000	1	1
Copenhagen	435	81,888	1	½
Rio de Janeiro	253	97,040	½	½
Miscellaneous	2,881	534,569	4½	2½
Total	57,697	20,312,093	100	100

'It thus appears that one of these meridians, that of Greenwich, is used by 72 per cent of the whole floating commerce of the world, while the remaining 28 per cent is divided among ten different initial meridians. If, then, the convenience of the greatest number alone should predominate, there can be no difficulty in a choice; but Greenwich is a national meridian . . .'

Fleming then went on to advocate his 1879 proposals that Greenwich plus 180° should be the zero for longitude and time. Passing through the Pacific, he said this would have all the advantages of Greenwich but could be thought of as being neutral. Fleming's proposal was not adopted but the table he quoted had a great influence on the final decision.

After Spain had said she would vote for Greenwich if Britain and the USA would adopt the metric system, General Strachey announced that Britain had just applied to join the international metric convention. However, although there was no legal bar to its being used in Britain for any purpose – and it was already being used extensively by scientists – the Government did not expect its use would ever be made compulsory. Sir William Thomson (later Lord Kelvin), the distinguished British physicist invited to address the conference though not a delegate, summed up the feelings of many of the delegates: '. . . but it seems to me that England is making a sacrifice in *not* adopting the metrical system. . . It cannot be said that one meridian is more scientific than another, but it can be said that one meridian is more convenient for practical purposes than another, and I think that this may be said pre-eminently of the meridian of Greenwich . . .'

So at last the vote on Resolution II was taken, in the afternoon of 13 October 1884. There were 22 ayes, 1 no (San Domingo) and 2 abstentions (France and Brazil). The conference had chosen Greenwich as Prime Meridian of the world – though it must be made clear that this was merely a recommendation to the respective governments, not an absolute commitment. During the final session, San Domingo explained her negative vote. She '. . . was bound to regard equity alone on the occurrence of the disagreement produced by the proposal of the Delegates of France, a nation renowned for being one of the first in intellectual progress. . . She was glad another French proposal was accepted almost unanimously [presumably Resolution VII about the decimal division of the circle] which was a good omen for future unanimous agreement on behalf of the general interest of science. '. . . That day will be saluted with a cordial *hosanna* by the Republic of San Domingo, which is always ready freely to give its assent to the progress of civilisation.'

III

'That from this meridian longitude shall be counted in two directions up to 180 degrees, east longitude being plus and west longitude minus.'

Discussion on this resolution – on how to express longitude – originally proposed by the USA, was longer than might have been expected for two reasons. First, it was considered fundamental in connection with the reckoning of time, as dealt with in subsequent resolutions; secondly, the Rome conference had proposed a 360° notation from west to east, but there was a strong school of thought that considered it should be counted from east to west, while an even stronger lobby considered that the current practice of counting 180° both ways should not be changed.

Somewhat inconsequentially, discussion started with the presentation by the US General Railway Time Convention of a plea that the Meridian Conference should not do anything to disturb the standard time system which had proved so satisfactory since its introduction the previous year. Sweden proposed to count longitude in one direction from east to west; Spain from west to east; Great Britain, thinking largely of the convenience of navigators, asked, 'Why change current practice?' Sandford Fleming, however, differed from his British colleagues, agreeing with Sweden and then going on to make a very long speech advocating his own Cosmic Time ideas based on the anti-meridian of Greenwich (see p. 129 above). Eventually, on the proposal of the USA, the resolution quoted above was adopted by 14 votes to 5, with 6 abstaining. The way the voting went can be seen in Table II (p. 143).

IV

'That the Conference proposes the adoption of a universal day for all purposes for which it may be found convenient, and which shall not interfere with the use of local or other standard time where desirable.'

The conference now turned from considering longitude to considering time, Resolution IV being devoted to general principles. Discussion on this was dominated by a long speech by W. F. Allen, Secretary of the US Railway Time Convention, extolling the virtues of the standard-time system (which he had personally done so much to promote) for use in ordinary life, leaving universal time for science and international telegraphs. Replying to those who said exact local time was essential, Allen pointed out that, for domestic purposes, several countries had, by adopting a standard time, already proved this not to be so. Great Britain had, said Allen (but see p. 114 above), kept Greenwich time since 13 January 1848, which differed from local mean time by about 8 minutes in the east of the kingdom and 22½ in the west; Sweden had kept the time of the fifteenth meridian east of Greenwich since 1 January 1879 with east and west difference of 36½ and 16 minutes respectively; the United States and Canada had adopted an hour-zone system on 18 November 1883 whereby cities like Portland (Me.), Atlanta (Ga.), Omaha (Neb.), and Houston (Tex.) all now kept a time – without any ill effects or inconvenience – which differed twenty minutes or so from local mean time. 'Nearly eighty-five per cent of the total number of cities in the United States of over ten thousand inhabitants have adopted the new standard time for all purposes, and it is used upon ninety-seven and a half per cent of all the miles of railway lines.' The resolution was adopted by 23 to nil, Germany and San Domingo abstaining.

V

'That this universal day is to be a mean solar day; is to begin for all the world at the moment of mean midnight of the initial meridian, coinciding with the beginning of the civil day and date of that meridian; and is to be counted from zero up to twenty-four hours.'

After a diversion to listen to various letters that had been sent to the conference, proposing diverse prime meridians such as Bethlehem, the frontier between Russia and the USA, the Great Pyramid and Le Havre (which is on the Greenwich meridian), the conference turned from the general principles of the proposed universal time to specific definitions, the main discussion centring on one question: should the Universal Day start at noon – as was the practice of astronomers and, for some nations, at sea – or conform to civil practice and start at midnight? One of the principal uses for the new universal time was for astronomy, so it seemed reasonable that astronomers should have to make no change in their practice – and make no break in their chronologies, which had been established in the time of Hipparchos – and the scientists at the Rome conference had so recommended. On the other hand, the other likely users, the operators of the worldwide system of electric telegraphs, must start their day at midnight to conform to everyday usage.

Adams (GB), though himself an astronomer, argued strongly that astronomers should conform to the civil day and that the Universal Day should start at midnight: noon was, after all, midday, or *midi*, not the beginning or end of the day. The advantages of retaining astronomical tradition on this point, he said, seemed small compared with the overwhelming disadvantages which would be imposed by having two conflicting methods of reckoning dates. 'If this diversity is to disappear, it is plain that it is the astronomers who will have to yield. They are few in number compared with the rest of the world. They are intelligent, and could make the required change without any difficulty, and with slight or no inconvenience', a sentiment not shared by the majority of his astronomical colleagues, as we shall see.

A Swedish resolution, supported by Austria-Hungary, Italy, the Netherlands, Switzerland and Turkey, in favour of starting the day at noon, was lost by 6 to 14, France, Germany, San Domingo and Spain abstaining. The Turkish delegate pointed out that, whatever the conference might recommend, in Turkey there would always be two times kept, *l'heure à la franque* from midnight to midnight, and *l'heure à la turque* from sundown to sundown, used by agricultural workers, from which Muslim prayers were counted.

The resolution quoted above, for the Universal Day to start at midnight on the prime meridian, was eventually passed by 15 to 2 – Austria-Hungary and Spain – with 7 abstentions, reversing the proposals of the Rome conference. An anomaly which was not discussed in 1884 (nor apparently in 1900) has raised its head more than a hundred years later in connection with the new millennium. According to Resolution V, the twenty-first century should begin 'for all the world' at 00h 00m 00s GMT (within a second of UTC) on 1 January 2001. There is no doubt that, by Universal Time, the new century will begin then, but the New Zealanders say that *their* new century will definitely begin 12 hours earlier (or thirteen because of Daylight Saving Time). We do not intend to take sides!

VI

'That the Conference expresses the hope that as soon as may be practicable the astronomical and nautical days will be arranged everywhere to begin at mean midnight.'

We have already seen how astronomers (following Hipparchos) used a 24-hour-clock system and (until 1925) *began* their day with the Sun's culmination at noon: this was convenient because it meant that the date did not change in the middle of a night's observing, as it would if civil reckoning had been used. At sea until the nineteenth century, on the other hand, the day was considered to *end* at noon on the civil day concerned. The log was kept in ship's time, that is local apparent time, adjusted at intervals for change of longitude: and by old practice the day would end (and the time be adjusted) at noon when the officer taking the latitude sight called 'Twelve o'clock, sir', and the Captain would reply 'Make it so!' Thus, noon marked the beginning of the day – say Monday – in astronomical reckoning, the middle ('midday') in civil reckoning, and the end of the day in nautical reckoning. While 6 a.m. Monday civil reckoning was also 6 a.m. Monday to the navigator, it was 18.00 *Sunday* to the astronomer. On the other hand, twelve hours later, at 6 p.m. Monday civil reckoning, it was 6 p.m. *Tuesday* to the navigator but 6.00 Monday to the astronomer. All of which was further confused because, when he entered harbour, the navigator (and the ship's log) reverted to civil reckoning. None of this mattered greatly until the *Nautical Almanac* (which followed astronomical practice) began to be used at sea, after which the opportunities for errors began to multiply. For example, the historian of, say, Cook's voyages has to be careful today because the journals of Cook himself and of his astronomer William Wales

Table II Voting at the International Meridian Conference, Washington, October 1884

Resolutions

	I Principle of Single Prime Meridian	II Greenwich Meridian	III 18°E and 180°W	IV Principle of Universal Day	V Solar Mean midnight 0 - 24h	VI Astronomical and Nautical days	VII Principle of Decimal angles and time
Page no. in this text	134	135	137	138	139	140	141
Austria-Hungary	AYE	AYE	ABSTAIN	AYE	NO		AYE
*Brazil	AYE	ABSTAIN	ABSTAIN	AYE	AYE		AYE
*Chile	NP	AYE	AYE	AYE	AYE		AYE
*Colombia	AYE	AYE	AYE	AYE	AYE		AYE
*Costa Rica	AYE	AYE	AYE	AYE	AYE		AYE
Denmark	NP	NP	NP	NP	NP		NP
*France	AYE	ABSTAIN	ABSTAIN	AYE	ABSTAIN		AYE
Germany	AYE	AYE	ABSTAIN	ABSTAIN	ABSTAIN		ABSTAIN
*Great Britain	AYE	AYE	AYE	AYE	AYE		AYE
*Guatemala	AYE	AYE	AYE	AYE	AYE		ABSTAIN
*Hawaii	AYE	AYE	AYE	AYE	AYE		AYE
Italy	AYE	AYE	NO	AYE	ABSTAIN	CARRIED	AYE
*Japan	AYE	AYE	AYE	AYE	AYE	WITHOUT	AYE
Liberia	NP	AYE	AYE	AYE	AYE	DIVISION	AYE
*Mexico	AYE	AYE	AYE	AYE	AYE		AYE
Netherlands	NP	AYE	NO	AYE	ABSTAIN		AYE
Paraguay	AYE	AYE	AYE	AYE	AYE		AYE
*Russia	AYE	AYE	AYE	AYE	AYE		AYE
San Domingo	AYE	NO	ABSTAIN	ABSTAIN	ABSTAIN		AYE
Salvador	AYE	AYE	AYE	AYE	NP		NP
*Spain	AYE	AYE	NO	AYE	NO		AYE
Sweden (with Norway)	AYE	AYE	NO	AYE	ABSTAIN		ABSTAIN
Switzerland	AYE	AYE	NO	AYE	ABSTAIN		AYE
Turkey	NP	AYE	ABSTAIN	AYE	NO†		AYE
*United States	AYE	AYE	AYE	AYE	AYE		AYE
Venezuela	AYE	AYE	AYE	AYE	AYE		AYE
AYES	21	22	14	23	15	–	21
NOES	0	1	5	0	2	–	0
ABSTAIN	0	2	6	2	7	–	3
Nos. of pages of discussion in official report	5	71	35	12	36	1	6

* Countries with scientific representatives
NP = Not present
† Changed from Aye to No in final session

each used a different reckoning for recording the same events, and these differed again at sea and in harbour.[36]

At the time of the Washington conference the Astronomical Day was very much alive but the Nautical Day had been abandoned in favour of civil reckoning by seamen of many nations. In the Royal Navy the demise of the Nautical Day occurred with the issue of an Admiralty Instruction dated 11 October 1805 (ten days before the Battle of Trafalgar) describing a new form of log-book 'to be kept in all King's ships . . . calendar or civil day is to be made use of, beginning at midnight. . . . It is necessary to remark to you, that the private night signal for each day of the month is to continue in force until day-light of the following day'[37] – the last of these observations was, of course, for exactly the same reason that astronomers preferred a day beginning at noon. Apart from the King's ships, the Nautical Day was abandoned by the East India Company in the 1820s but remained in use in many other merchant ships until well into the middle of the nineteenth century. As was made evident at the conference, however, the Nautical Day was still in use in ships of some nations at that time.

However, there had been so much debate on the matter of the time at which the Universal Day should start for the previous resolution that Resolution VI was carried with almost no discussion and without a division. Nevertheless, despite the apparent unanimity, it would be thirty-five years before the Astronomical Day problem was resolved.

VII

'That the Conference expresses the hope that the technical studies designed to regulate and extend the application of the decimal system to the division of angular space and of time shall be resumed, so as to permit the extension of this application to all cases in which it presents real advantages.'

This resolution, following a similar one in the Rome conference, was proposed by France. After some discussion as to whether the subject of decimalization of angles and time was not outside the conference's terms of reference, it was put to the vote and adopted by 21 to 0, with Germany, Guatemala and Sweden abstaining.

Before the conference ended, Great Britain proposed two more resolutions, one recommending standard time for local civil time 'at successive meridians distributed around the earth, at time-intervals of either ten minutes, or some integral multiple of ten minutes, from the prime meridian'. (The Swedish astronomer Gyldén had proposed a standard time system based on time intervals of 2½, or 10 minutes.) The other resolution proposed that arrangements for the adoption of the Universal Day in international telegraphy should be left to the International Telegraph Congress. However, it was decided that the questions raised were covered by resolutions already adopted, so these last two were withdrawn.

And there, except for a summing up in the Final Act and some short speeches of thanks on 1 November, ended the International Meridian Conference of 1884, recommending (among other things) that the Universal Time should be Greenwich Time. It is interesting that the three resolutions recommending matters of principle – the desirability of having a single prime meridian, a Universal Day and decimal angle and time – were carried almost unanimously. Of the three specific resolutions defining the prime meridian and universal time, Great Britain and the USA voted in favour, and were followed by a substantial number of countries. On the other side, Brazil, France and San Domingo abstained or voted against. Austria-Hungary, Germany, Italy, the Netherlands, Spain, Sweden, Switzerland and Turkey then supported the UK/USA bloc on the choice of the Greenwich meridian, but abstained or voted against on the other issues.

The civil, astronomical, and nautical days. The civil day has always started at midnight, but from the time of Hipparchos until 1925, astronomers started their day at noon, using the 24-hour clock. At sea, however, the day was considered to end at noon; so the nautical day was twenty-four hours behind the astronomical day.

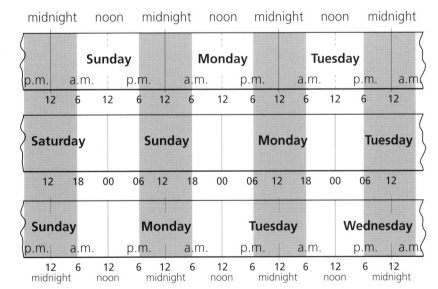

Civil reckoning
Day ends at midnight

Astronomical reckoning
Day begins at noon

Nautical reckoning
Day ends at noon

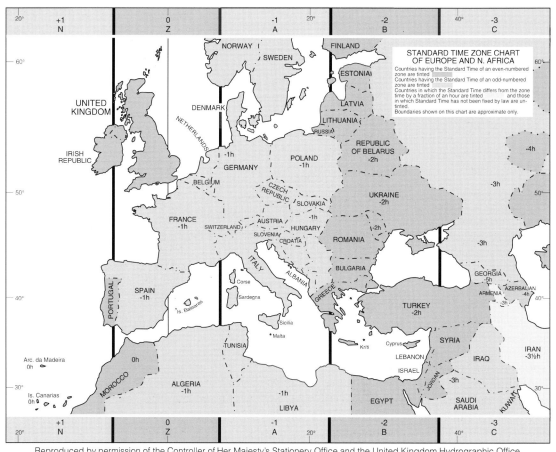

STANDARD TIME ZONE CHART
OF EUROPE AND N. AFRICA

Countries having the Standard Time of an even-numbered zone are tinted
Countries having the Standard Time of an odd-numbered zone are tinted
Countries in which the Standard Time differs from the zone time by a fraction of an hour are tinted   and those in which Standard Time has not been fixed by law are un-tinted.
Boundaries shown on this chart are approximate only.

Greenwich time for the world

1884 - 1939

Standard time

The principal impact of the Washington conference on the man-in-the-street was the adoption, country by country, of a new system of time zones based upon the world's new Prime Meridian, Greenwich – and this despite the fact that the time-zone system, though discussed, was not specifically recommended by the conference.

At first, many of those who conceived the Universal Day and Universal Time – or Terrestrial Time, or Cosmic Time – envisaged one single time being used all over the world for all purposes. But this view never prevailed beyond one or two enthusiasts. However, Dowd's time-zone system with one-hour differences between zones seemed to many to be the best possible compromise between Universal Time and local time: no one had to keep his clock much more than 30 minutes different from local time, and yet the minute hands were all the same, only the hour hands differing from zone to zone. To those who complained that the Sun was no longer precisely on the meridian at noon, it was pointed out that the change from Apparent Time to Mean Time – by then a *fait accompli* in civilized countries the world over – was an even more fundamental change than that from local mean time to Standard Time. How many people realize that, as a result of the change from apparent to mean time, afternoons in November are almost half an hour shorter than the mornings, while in February the mornings are half an hour shorter than the afternoons?

While the conference was meeting, four countries were already keeping the new system – Great Britain, Sweden, the USA and Canada. As can be seen in Table III, a standard time based on the Greenwich meridian was adopted country by country, often first for railway and telegraph purposes, followed soon for legal and general purposes. By 1905 the only major countries not conforming were France, Portugal, Holland, Greece, Turkey, Russia, Ireland (oddly enough) and most of Central and South America except Chile. Of thirty-six nations then using Standard Time, twenty had adopted Greenwich as the basis of their system; of the remaining sixteen, no two agreed. France did not conform at first but, with Algeria,

Time-zone chart, 1996, showing Europe, the Middle East and North Africa (opposite) and the world (following pages). Zones are 15° (= 1 hour) wide. The zone numbers (-12, +12, etc.) at the top and bottom indicate the correction in hours to be applied to zone time to obtain GMT, or Universal Time (UTC): e.g. 14.00 (-10) in Sydney, Australia, is the equivalent of 04.00 GMT. Zone letters, Z (zulu time) = GMT, R = +5, for example, are used to indicate the zone time being used in international communications.

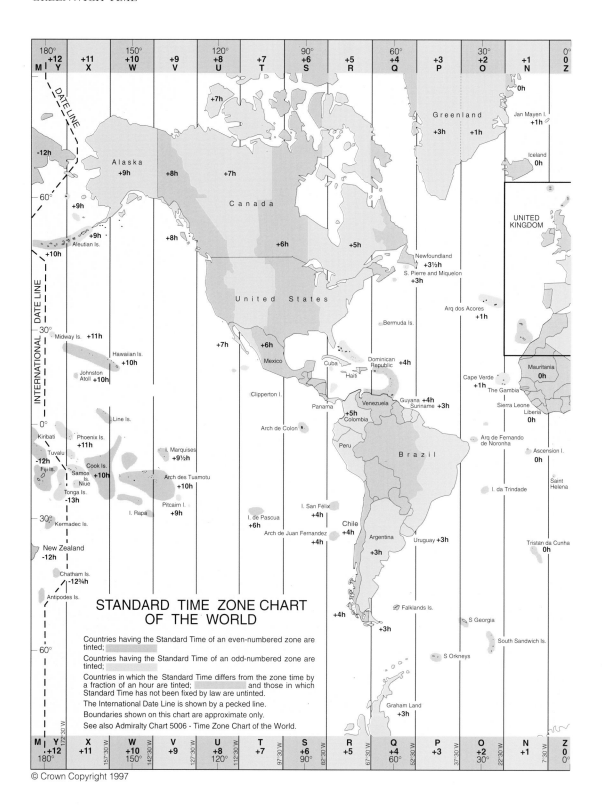

STANDARD TIME ZONE CHART OF THE WORLD

Countries having the Standard Time of an even-numbered zone are tinted;

Countries having the Standard Time of an odd-numbered zone are tinted;

Countries in which the Standard Time differs from the zone time by a fraction of an hour are tinted; and those in which Standard Time has not been fixed by law are untinted.

The International Date Line is shown by a pecked line.

Boundaries shown on this chart are approximate only.

See also Admiralty Chart 5006 - Time Zone Chart of the World.

© Crown Copyright 1997

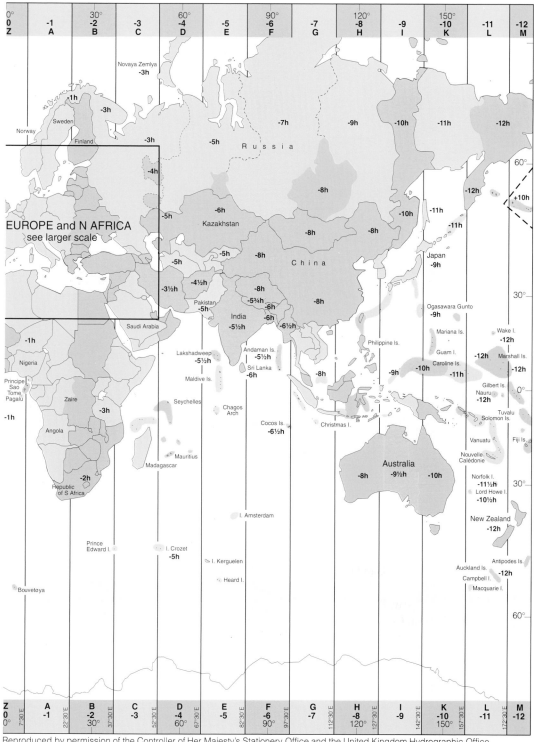

Reproduced by permission of the Controller of Her Majesty's Stationery Office and the United Kingdom Hydrographic Office.

Table III Dates of adoption of zone times based on the Greenwich meridian, including half-hour zones

N.B.   (1) The names of countries given here are those in general use on the date quoted.

(2) In some cases, standard times were brought into use for railways and telegraphs before the dates stated.

(3) Countries with no sea coast are, in general, omitted.

1848	Great Britain (legal in 1880)
1883	Canada, USA (legal in 1918)
1884	Serbia
1888	Japan
1892	Belgium, Holland*, S. Africa except Natal
1893	Italy, Germany, Austria-Hungary (railways)
1894	Bulgaria, Denmark, Norway, Switzerland, Romania, Turkey (railways)
1895	Australia, New Zealand, Natal
1896	Formosa
1899	Puerto Rico, Philippines
1900	Sweden, Egypt, Alaska
1901	Spain
1902	Mozambique, Rhodesia
1903	Ts'intao, Tientsin
1904	China Coast, Korea, Manchuria, N. Borneo
1905	Chile
1906	India (except Calcutta), Ceylon, Seychelles
1907	Mauritius, Chagos
1908	Faroe Is., Iceland
1911	France, Algeria, Tunis and many French overseas possessions, British West Indies
1912	Portugal and overseas possessions, other French possessions, Samoa, Hawaii, Midway and Guam, Timor, Bismarck Arch., Jamaica, Bahamas Is.
1913	British Honduras, Dahomey
1914	Albania, Brazil, Colombia
1916	Greece, Ireland, Poland, Turkey
1917	Iraq, Palestine
1918	Guatemala, Panama, Gambia, Gold Coast
1919	Latvia, Nigeria
1920	Argentine, Uruguay, Burma, Siam
1921	Finland, Estonia, Costa Rica
1922	Mexico
1924	Java, USSR
1925	Cuba
1928	China Island
1930	Bermuda
1931	Paraguay
1932	Barbados, Bolivia, Dutch East Indies

*Legal time reverted to Amsterdam time 1909; to Central European Time 1940.

1934	Nicaragua, E. Niger
By 1936	Labrador, Norfolk I.
By 1937	Cayman Is., Curaçao, Ecuador, Newfoundland
By 1939	Fernando Po, Persia
1940	Holland
By 1940	Lord Howe I.
By 1948	Aden, Ascension I., Bahrein, British Somaliland, Calcutta, Dutch Guiana, Kenya, Federated Malay States, Oman, Straits Settlements, St Helena, Uganda, Zanzibar
By 1953	Raratonga, South Georgia
By 1954	Cook Is.
By 1959	Maldive I. Republic
By 1961	Friendly Is., Tonga Is.
By 1962	Saudi Arabia
By 1964	Niue Is.
1972	Liberia

In 1978, Guyana was keeping + 3h 45m; Nepal - 5h 45m; Chatham Island - 12h 45m. Otherwise, all countries were keeping time within an even hour or half-hour of Greenwich.

Principal sources

Koppenstatter (ed.), *Zonen, und Sommerzeiten aller Länder und Städte den Erde* (München [1937]).

US National Bureau of Standards. 'Standard Time throughout the World', *Circular of the Bureau of Standards*, no.399 (15 Sept. 1932).

The Observatory, Feb. 1901, 88-91.

Abridged Nautical Almanac annually.

adopted Paris Mean Time as *l'heure nationale* by the law of 4 March 1891. As had happened in many other countries, this was largely a legalization of the railway time already being generally used throughout the country. In December 1891 F. Pasquier wrote in *Ciel et Terre* that 'the almost unanimous adhesion of civilized nations to the meridian of Greenwich should cause the partisans of other meridians to lay down their weapons, and henceforth all efforts should be directed to settling definitely an hourly unification at once simple, rational and practical.'[1]

Some five years later, on 27 October 1896, Deputy Deville introduced a Bill into the Chamber of Deputies proposing Greenwich Mean Time as the legal time in France. The Bill, with the amendment that this should be expressed as Paris Mean Time diminished by 9 minutes 21 seconds (which comes to the same thing as GMT but avoids the use of the word Greenwich), was passed on 24 February 1898 by the Chamber to the Senate, who duly passed it to a Commission – and there it stayed for twelve long years. Apparently the Ministries of Commerce, Industry, Posts and Telegraphs, and Public Works all accepted the Bill but it was strongly opposed by the Ministries of Public Instruction and by the Navy.

Eventually, on 9 March 1911 (by which time radio had become a reality), the following law was passed and came into effect on the night of 10-11 March: 'Legal time in France and Algeria is Paris Mean Time, retarded by 9 minutes 21 seconds.'[2] This law remained in force until 9 August 1978 when it was repealed by a decree which stated that henceforward French legal time should be determined from Coordinated Universal Time (UTC, which by definition is kept within 0.9 seconds of GMT), legal time being obtained by adding or subtracting an exact number of hours to UTC. The significance of this decree is discussed in Chapter 7.[3]

Of the other countries mentioned above, Portugal adopted a time based on the Greenwich meridian in 1912, Brazil and Colombia in 1914, Greece, Ireland, Poland and Turkey in 1916, Argentina and Uruguay in 1920, the USSR in 1924. Holland, whose railways had kept GMT since 1892 but who had used Amsterdam mean time instead of local time for other purposes since 1909, was forced to adopt Central European Time on 19 May 1940 during the German Occupation, a usage which was finally confirmed in 1956. The last major nation to conform was Liberia, whose legal time remained 44m 30s slow on GMT until January 1972 when, in honour of the President's birthday, she adopted GMT.

Unification of civil and astronomical days

Astronomers, however, were more conservative. Despite Resolution VI of the Washington conference being passed without a division, there was very considerable opposition from certain parts of the astronomical world to the idea of the astronomical day starting at midnight instead of noon. This was voiced volubly at an astronomical congress in Geneva the following year when the Washington Resolution was condemned by the majority of the astronomers present, particularly Newcomb (USA), Auwers (Berlin), Gyldén (Stockholm), and Tietjen (Berlin), though defended by O. Struve (Pulkowa). In passing, it is worth noting that, as long ago as 1804, Laplace had proposed this unification and, after a long discussion, his view had been adopted by the Bureau des Longitudes by 7 votes to 5, though the *Connaissance des Temps* remained faithful to the old notation.[4]

In 1885 Britain adopted the Civil Day for spectroscopic, photographic, magnetic and meteorological observations, while the 24-hour clocks in Greenwich Observatory (including the public clock outside the gates) were, on 1 January 1885, set to civil time, with the day starting at midnight. However, the astronomical observations and the *Nautical Almanac* remained in astronomical time, pending general international agreement. In 1893 the Joint Committee of the Canadian Institute and the Astronomical and Physical Society of Toronto, under the chairmanship of Sandford Fleming, sent a circular to astronomers of all nations, asking the question: 'Is it desirable, all interests considered, that on and after the first day of January, 1901, the astronomical day should everywhere begin at Mean Midnight?'[5] Of the 171 replies received, 108 were in favour, 63 against. Broken down into nationalities, 18 countries were in favour (including the United States, though Simon Newcomb did not reply), 4 against (Germany, Holland, Norway, Portugal).

About the same time W. M. Greenwood of Glasson Dock, Lancaster, England, sent a circular to shipmasters of all nations asking four questions broadly covering the main recommendations of the Washington conference. Of 409 replies received to Question 3 on the matter of midnight *v.* noon, 399 chose the former, 10 the latter. On the 24-hour-clock system, there were 22 against.[6] In France the Bureau des Longitudes, when asked by the Ministry of Foreign Affairs in 1884 to give an opinion, voted 7 to 5 in favour of the change.[7] But, despite these opinions, no general agreement was reached among astronomers and the nautical almanacs continued to use noon to begin their day. Then in 1917, at the Anglo-French Conference on Time-keeping at Sea, this resolution was passed:

> 11. That, from the point of view of seamen it would be a considerable advantage if a day commencing at 0 hours midnight were substituted for the astronomical day commencing at 0 hours noon in all nautical publications, and that the Royal Astronomical Society should be asked to ascertain the views of astronomers as to such an alteration, including possibly the general substitution of the civil for the astronomical day.[8]

Although there were dissenting views, the replies to the RAS's circular were generally favourable to the change, many astronomers grudgingly agreeing that, if it would help seamen, it should be done. So, from the issue for 1925, the British and many other nautical almanacs began to use the civil day, beginning at midnight.

However, the arguments continued among astronomers, largely centred on nomenclature: before 1925, the GMT day began at noon; should a time-scale differing 12 hours from this still be called by the same name? In America, Greenwich Civil Time (GCT) was favoured as the name for the new time scale; the British Admiralty disagreed because of the possibility of confusion with a warship's Gunnery Control Tower – GCT. Then, in 1928, the International Astronomical Union recommended internationally that the time scale used in almanacs should be called Universal Time (UT). Thereafter, UT began to be adopted for astronomy, though the term GMT is still used in navigational publications, in rail and air timetables, and for international cable and radio communications.[9]

The International Date Line

A corollary to the establishment of a prime meridian for longitude and time is that there should be an anti-prime meridian 180° away – an International Date Line, on either side of which the date is different. At one particular instant, it will be Monday to the west of that line but Sunday to the east of it. The problem of this change of date came to light as early as the first circumnavigation. When the survivors of Ferdinand Magellan's expedition reached civilization in 1522, having sailed through the Straits now called after him and then westwards across the Pacific to the Philippines (where Magellan himself was killed) and the Spice Islands, and then around the Cape of Good Hope, they discovered, not only that they had practical proof that the Earth was round, but also that they had somehow gained a day in their lives, as was explained by Antonio Pigafetta, the young Italian nobleman from whose diary we learn so much about the voyage. Despite the fact that the Cape Verde Islands were Portuguese and therefore enemy territory, the Spaniards in the one remaining ship, the *Victoria*, were forced to call there for supplies. According to Pigafetta's diary, it was Wednesday 9 July 1522, but,

> In order to see whether we had kept an exact account of the days, we charged those who went ashore to ask what day of the week it was, and they were told by the Portuguese inhabitants of the island that it was Thursday, which was a great cause of wondering to us, since with us it was only Wednesday. We could not persuade ourselves that we were mistaken; and I was more surprised than the others, since having always been in good health, I had every day, without intermission, written down the day that was current. But we were afterwards advised that there was no error on our part, since as we had always sailed towards the west, following the course of the sun, and had returned to the same place, we must have gained twenty-four hours, as is clear to any one who reflects upon it.[10]

More than 150 years later, the English circumnavigator William Dampier (1652-1715) came across another aspect of the problem. Having travelled westwards around the world (as Magellan had done), Dampier's ship reached the Philippines on 14 January 1687:

> It was during our stay at *Mindanao*, that we were first made sensible of the change of time, in the course of our Voyage. For having travell'd so far Westward, keeping the same Course with the Sun, we must consequently have gain'd something insensibly in the length of the particular Days, but have lost in the tale, the bulk, or number of the Days or Hours. According to the different Longitudes of *England* and *Mindanao*, this Isle being West from the *Lizzard*, by common Computation, about 210 Degrees, the difference of time at our Arrival at *Mindanao* ought to be about 14 Hours: And so much we should have anticipated our reckoning, having gained it by bearing the Sun company. Now the natural Day in every particular place must be

consonant to itself: But this going about with, or against the Sun's course, will of necessity make a difference in the Calculation of the civil Day between any two places. Accordingly, at *Mindanao*, and all other places in the *East Indies*, we found them reckoning a Day before us, both Natives and *Europeans*; for the *Europeans* coming eastward by the Cape of *Good Hope*, in a Course contrary to the Sun and us, where-ever we met they were a full Day before us in their Accounts. So among the *Indian Mohometans* here, their *Friday*, the Day of their Sultan's going to their Mosques, was *Thursday* with us; though it were *Friday* also with those who came eastward from *Europe*. Yet at the *Ladrone* Islands, we found the *Spaniards* of *Guam* keeping the same Computation with ourselves; the reason of which I take to be, that they settled that Colony by a Course westward from *Spain*; the *Spaniards* going first to America, and thence to the *Ladrones* and *Philippines*. But how the reckoning was at *Manila*, and the rest of the *Spanish* Colonies in the *Philippine* Islands, I know not; whether they keep it as they brought it, or corrected it by the Accounts of the Natives, and of the *Portugueze, Dutch and English*, coming the contrary way from *Europe*.[11]

Thus there was a discrepancy as to date around the borders of the Pacific according to whether the colonizers came from the east or the west. The Portuguese, then the Dutch, the French and the British, came to the East Indies by way of the Cape of Good Hope. The Spaniards, on the other hand, reached the Philippines and Ladrones from America. Until 1844 the Philippines kept 'American date' while Celebes, in the same longitude, kept 'Asiatic date'.

To the north there was another example of 'reaching into the wrong hemisphere', this time with Asia reaching into America. Inspired by Bering's discoveries, Russian fur-traders settled in Alaska as early as 1745. In course of time Alaska became a Russian colony, in which the Orthodox Church, with its Julian calendar, was established, the dates being the same as those in St Petersburg and Moscow. In the nineteenth century American traders streamed into Alaska causing worry to the Orthodox priests, not only by their adherence to the Gregorian calendar (which affected the day of the month) but also because they insisted upon observing their day of rest on the day that the Russians claimed to be a Monday. Eventually, in 1867, the Territory was bought by the United States for seven million dollars, and the Gregorian calendar and American date introduced.

In 1879 the British governor of the Fiji Islands (through which the 180° meridian passes) enacted an ordinance to say that all the islands should keep the same time, Antipodean Time. The King of Samoa, however, under pressure from American business interests, went the other way, changing the date in his kingdom from the Antipodean to the American system, ordaining – by a masterpiece of diplomatic flattery – that the Fourth of July should be celebrated twice in that year.[12] The date line as originally drawn had a kink to the westward of the

Hawaiian Islands to include Morrell and Byers Islands which appeared on nineteenth-century charts at the western end of the Hawaiian chain. It was then proved that they did not exist, so the date line was straightened out. The Cook Islands remain on a different side of the date line from New Zealand, by whom they are administered. There is a saying in Raratonga: 'When it's today in Raratonga, it's tomorrow in Wellington.'

It is worth noting that the drawing of the International Date Line is not the result of a formal international agreement but, in the words of the Hydrographer of the Navy in Britain, is 'merely a method of expressing graphically . . . the differences of date which exist among some of the island groups in the Pacific'.

Radio time signals

The radio time signal was a fundamental step in the development of the dissemination of time, particularly for navigational purposes. At last it was possible for a ship to check her chronometers out of sight of land. The radio time signal also drove the final nails into the coffin of the lunar-distance method of finding longitude at sea. With chronometers cheap enough to be carried in ships of every size, lunar distances had long been superseded for day-to-day use, though the ability to use them occasionally had nevertheless to be retained so that the navigator could check his chronometers when other means were not available. Radio removed the need for this and the British *Nautical Almanac* ceased to publish lunar-distance tables in 1907, though instructions on how to compute and reduce them continued until 1924. One of the most important dates in our story was therefore 29 March 1899 when Guglielmo Marconi, from near Boulogne with an apparatus invented by the Frenchman Edouard Branly, detected a signal sent over the English Channel from near Dover (all of 45 km distant). So wireless telegraphy, soon to be called radio-telegraphy or just plain radio, proved to be a practical means of communication, a proof reinforced when, two years later, Marconi was in Newfoundland to receive the first transatlantic radio message.

The earliest wireless time signals for navigational purposes seem to have been those broadcast by the US Navy on low power from Navesink, NJ, in the spring of 1904, leading to the first regular transmissions every day at noon EST from Washington, DC, from January 1905. High-power radio time signals from Arlington, Va., began in December 1912.[13] In Germany, experimental transmissions began from Norddeich Radio (30 km north of Emden) in 1907, regular transmissions from May 1910.

France had never had a full-scale time service controlled by a national observatory as Britain had had since 1852, and it was not until 1880 that telegraphic time signals were sent to towns in France that wanted them, and then only once a week on Sundays, at first to Rouen and Le Havre, later to La Rochelle, Nancy, Saint-Nazaire, Chambéry and Cluses.[14] In 1908, however, the Bureau des Longitudes recommended that wireless time signals should be broadcast regularly from the Eiffel Tower, a project which the Ministry of War agreed to sponsor. All was ready by January 1910 but, just as the service was about to start, there was a particularly bad flooding of the Seine and underground installations on the Champ-de-Mars were inundated. Repairs were made and regular transmission of time signals started on 23 May 1910, daily at midnight Paris Mean Time. On 21 November a daytime signal began at 11.00 daily. On 9 March 1911, by the law of the same date already mentioned, France put her clocks back to the equivalent of GMT but the Eiffel Tower did not follow suit until 1 July when time signals started at 10.45 and 23.45 GMT daily. From February 1912 an additional rhythmic signal on the vernier principle was broadcast, allowing the error of a clock or watch to be found to an accuracy of a hundredth of a second.[15] Similar 'scientific' time signals were broadcast from many other countries after the Second World War; that from Rugby, England, from 1927 to 1958.

There were no early moves in England to institute wireless time signals, which would, of course, have been of particular value to shipping. This is surprising but perhaps it was thought that, in time of peace, the time-balls in various ports – and listening to foreign time signals – would be adequate, while, in time of war, radio time signals would be stopped anyway. Nevertheless, a wireless room was fitted up at Greenwich Observatory so that foreign signals could be received, compared with the time determined at Greenwich, and discrepancies reported to the observatory responsible.

The Bureau International de l'Heure

By 1911 it had been found that wireless time signals sent from the various stations could differ from each other by several seconds. Having removed one of the obstacles to international co-ordination in this matter by adopting GMT, the French took the initiative and, in May 1912, invited certain other governments to send delegates 'to study ways and means of effecting practical unification of radio time signals, and to prepare plans for an international time service to suit the needs of all'.[16]

The conference, with representatives from sixteen states, assembled at Paris Observatory on 12 October 1912, the directors of most national observatories being present: Britain sent her new Astronomer Royal, F. W. Dyson, and the Assistant Hydrographer, Capt. J. F. Parry; the USA sent Prof. Asaph Hall of the Naval

Observatory; Germany sent Prof. W. Foerster, former director of Berlin Observatory. After lengthy discussions, a formal proposal was made to establish an International Time Commission which would have three main objectives: to secure the unification of time signals; to secure the universal use of GMT; and to create an international organization to be called the Bureau International de l'Heure (BIH), with the task of co-ordinating results from observatories and deducing the most exact time – *l'heure définitive*. An interesting worldwide network of time signals was also proposed, to start on 1 July 1913, a target date which was to be met by few of the countries concerned:

Station	GMT of signal	Station	GMT of signal
Paris	0 h	Norddeich-Wilhelmshavn	12
Fernando I. (Brazil)	2	Fernando I. (Brazil)	16
Arlington (USA)	3	Arlington (USA)	17
Manila (Philippines)	4	Massawa (Eritrea)	18
Mogadishu (Somalia)	4	San Francisco (USA)	20
Timbuctu	6	Norddeich-Wilhelmshavn	22
Paris	10		

A second International Time Conference met in Paris in October of the following year. This was of a diplomatic rather than a scientific character, was attended by representatives of thirty-two states, and led to the drawing-up of statutes for a new body to be called the International Time Association whose principal purpose was to superintend the proposed BIH. At the request of a committee appointed by the conference, the director of Paris Observatory set up a provisional Time Bureau in 1913 in anticipation of the ratification by the various states of the conference proposals. In the event, the outbreak of the First World War less than a year later prevented full ratification. Despite many difficulties, however, the provisional BIH continued to operate throughout the war.

In 1918 the Royal Society in London took the initiative in a move to re-start the international scientific co-operation which had been developing before the war by calling the Inter-Allies Conference of Scientific Academies, which met in London in October 1918, in Paris in November 1918, and in Brussels in July 1919. At this last meeting, the International Astronomical Union (IAU) was formed, meeting the same year and immediately setting up a Time Commission within its own structure to serve the same purpose as the proposed Time Association of 1913, principally to supervise the BIH which was finally established on an international basis on 1 January 1920, operating from Paris Observatory, with Guillaume Bigourdan (1851-1952, who had given the opening address at the 1912 conference) as its first director.[17]

Daylight Saving Time

Daylight Saving Time, or Summer Time as it is known in Britain, was the brain-child of William Willett (1857-1915), a London builder living at Petts Wood in Kent. In a pamphlet circulated in 1907 to many Members of Parliament, town councils, businesses and other organizations, he pointed out 'that for nearly half the year the sun shines upon the land for several hours each day while we are asleep, and is rapidly nearing the horizon, having already passed its western limit, when we reach home after the work of the day is over. . .'[18] He proposed to improve health and happiness by advancing the clocks 20 minutes on each of four Sundays in April, and retarding them by the same amount on four Sundays in September. In addition to improvements in health and happiness – his main objective – Willett also claimed that, with electricity costing 1/10 of a penny per hour, the country would save £2½ million, even taking into account the loss of profit to producers of artificial light.

Though the scheme was ridiculed and met with considerable opposition, particularly from farming interests in England, nevertheless a Daylight Saving Bill was drafted in 1909 and introduced in Parliament several times, though it met with no success before war broke out.

In April 1916, however, Daylight Saving Time was introduced as a wartime measure of economy, not only in Britain but, within a week or so, in nearly all countries, both allied and enemy. Willett had died the previous year so never saw his ideas put into effect. Though many countries abandoned Daylight Saving Time immediately after the war, most reintroduced it eventually, and some even began to keep it throughout the year. In the Second World War Britain kept 'Summer Time' (BST) in the winter, with 'Double Summer Time' (DBST, two hours in advance of GMT) in the summer.

Between 1968 and 1971 Britain tried the experiment of keeping BST – to be called British Standard Time – throughout the year, largely for commercial reasons because Britain would then conform to the time kept by the rest of the countries in the European Community. This measure met with considerable opposition from the country generally – and the more so the further west and north. In the summer, no one minded; in the winter, however, the children of Glasgow and points north, for example, *always* had to go to school in the dark. The experiment was abandoned and, since 1972, Britain has kept GMT in winter and BST in summer. Most other countries now keep Daylight Saving Time in summer, some (like France) throughout the year. 'Spring forward, fall back' – so runs the only mnemonic the author knows for deciding that problem: 'which way should the clocks go?'

Standard time at sea

As we have seen, the standard time system based on the Greenwich meridian (sometimes called the zone-time system) quickly established itself ashore. At sea, there was at first no such agreement. Though the navigator used GMT for his calculations, for domestic purposes afloat it was the practice to put the clocks forward or back so as to make them show the exact apparent time of the vessel's noon position. In wartime, this was extraordinarily inconvenient. In a convoy, for example, the Commodore could not simply say: 'The convoy will alter course at noon.' He had to specify the longitude he was using – because every ship in the convoy was keeping a different noon.

In June 1917, therefore, an Anglo-French Conference on Timekeeping at Sea assembled in London. This recommended that the zone-time system should be used at sea, clock changes required by changes of longitude being made preferably in one-hour steps. This recommendation was immediately adopted by British and French ships, both naval and mercantile. Ships of most other nations soon followed suit so that, by a few years after the war, zone time was kept at sea, certainly by almost all naval ships and by many non-naval ships as well. Nevertheless, up to the Second World War, the old practice of changing ship's time at midday prevailed in many independent merchant ships.

Greenwich time in the home

As far as domestic time signals in Britain are concerned, GMT seems first to have come into the home by radio when the British Broadcasting Company (before it became a corporation) broadcast the chimes of Big Ben to usher in the year 1924. Late in 1923 Frank Dyson, the Astronomer Royal, had visited John Reith, Director General of the BBC, and discussed the idea of public time signals being broadcast. The famous 'six-pip' time signal – 'pips' to mark seconds 55, 56, 57, 58, 59, 60 – was Dyson's brain-child, devised in discussion with Frank Hope-Jones, inventor of the free pendulum clock, who had himself originally advocated a 'five-pip' signal.

On 5 February 1924 Dyson broadcast to the nation, inaugurating the new service. A little later he was presiding over a dinner of the British Horological Institute at which Hope-Jones was the guest of honour. Some wag, remembering the latter's connection with time signals, handed him six orange pips on a plate. With much ceremony, Hope-Jones formally presented the sixth pip to Dyson in the chair![19]

Wireless time signals in Britain in the home, 1923. Loudspeaker reception with Marconiphone V2.

The speaking clock

Time by telephone has a fairly long history. In 1905 Paris Observatory established a telephonic time signal available on demand, by placing a microphone in the mean-time clock case when needed, transmitting the clock ticks while at the same time a member of the observatory staff counted the minutes and seconds verbally on another instrument on the same line. In 1909 Hamburg Observatory established a slightly less precise telephone time service, also on demand.[20] The original Paris telephone time service proved very popular – but time-consuming for observatory staff. On 14 February 1933, therefore, a new service – *l'horloge parlante* – was offered by the Paris Observatory. This was a fully automatic 'speaking clock', available to any telephone subscriber who dialed the appropriate number, ODEon 84.00 (today 36-99). Similar systems were already in use in Strasbourg and also outside France.[21]

A similar 'speaking clock' was brought into use in Britain on 24 July 1936. On dialling TIM the subscriber would hear the time every ten seconds from the 'golden voice': 'At the third stroke, it will be six, fifty-seven, and twenty seconds.'

George Graham's 'no. 3' was the transit clock at Greenwich from 1750 to 1821; that is, the one used for stellar observations on Bradley's Prime Meridian when that was first established. From 1833 to about 1924 it was used to control the Greenwich time-ball, although the moment of drop was controlled by the Shepherd master-clock from 1852. Graham was the leading instrument maker of his time and an early supporter of John Harrison.

The meridian moved

The official maps of Great Britain are published by an organization known as the Ordnance Survey, the original triangulation for which took place between 1783 and 1853. Between 1936 and 1957 a retri-angulation of Great Britain was carried out. As part of this, observations were taken in 1949 to connect other stations of the survey with Airy's transit circle, which, must, by definition, be 00°00′00″ in longitude. In fact i transpired that, even after th angles had been re-observed, th retriangulation gave a longitud value for Airy's trans circle of 00°00′00″.418 east Greenwich – which should l impossible. Consternation! Th discrepancy in longitude which w the equivalent of 8.04 metres (26.39 feet) on the ground, was f greater than might be expected fro errors of observation, even in th eighteenth century. There was alsc discrepancy in latitude, amountin to 0.039 seconds of arc (= 3.95 1.21 m), but this was acceptable.

The person who provided t answer to the puzzle was the Royal Observatory's then Chief Assistant, Dr. R. d'E. Atkinson. He pointed out that, when General Roy had carried out the Principal Triangulation in 1787, the great theodolite had been erected immediately over Bradley's transit instrument which then defined the Greenwich meridian. Pond's transit instrument replaced

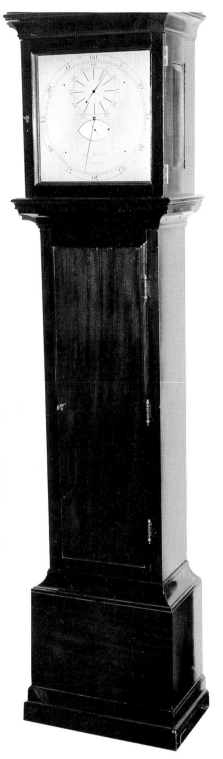

Bradley's in 1816, but it was mounted on the same piers, so there was no change in the meridian. However, in the late 1840s Airy decided that a new and much larger meridian instrument was needed. So that there should be no break in the sequence of transit observations, the new transit circle was erected in the old Circle Room, east of Bradley's Transit Room, while regular transit observations – particularly necessary for time determination, for example – continued on Pond's transit instrument.

On the first observing day in the new half-century, 4 January 1851 (the methodical Airy would

Sir George Biddell Airy (1801-92), seventh Astronomer Royal, painted by John Collier in 1884, the year that Greenwich was accepted as marking the world's Prime Meridian of Longitude.

have liked this to have been 1 January, but English weather frustrated him), the new transit circle was brought into use – and this had the effect of moving the Greenwich Meridian some 19 feet to the eastward, a difference of less than $\frac{1}{50}$ second in the time of transit, a quantity which at that date was too small to be measurable. When after 1884 Greenwich was chosen as the world's Prime Meridian, one country which did not have to change its maps was Britain – or so one would assume. However, it seems that, although Airy did inform the Ordnance Survey of the proposed change in instruments in 1850, they failed to change the records.

When all this was taken into account, the discrepancy between the old and the new triangulations was reduced to 6 centimetres in latitude and 1.95 metres in longitude.[22]

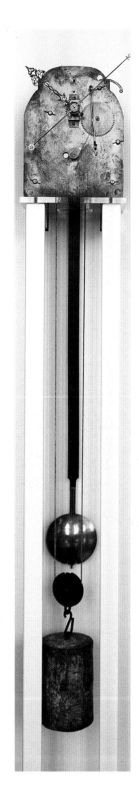

SEVEN

A clock more accurate than the Earth

Bᴇᴄᴀᴜsᴇ this story is primarily that of the distribution and the uses made of time, we have so far made but passing reference to observatory clocks. Until recently the fundamental time-keeper was the rotating Earth, and time was found at frequent intervals by astronomical observations; clocks were only used to 'keep' time in the comparatively short intervals between observations. In this chapter, however, it is the developments in the clocks themselves – and the consequences of those developments – that form the main part of the story because, within the last fifty years, man-made clocks have been developed which are better timekeepers than the Earth itself.

In the first two hundred years of the Royal Observatory's existence there was some increase in the precision of the pendulum clock, due to the invention of the dead-beat escapement and the temperature-compensated pendulum by Graham and others early in the eighteenth century, but these were not really fundamental developments. In 1676 Flamsteed's Great Clocks could be relied upon to about 7 seconds per day; in 1870 Airy's barometrically compensated regulator clock (Dent No.1906) was accurate to about 0.1 seconds per day (in accuracy, somewhat ahead of its time).

In the last decade of the nineteenth century the leading astronomical observatories (though not Greenwich) began acquiring clocks significantly more accurate than their predecessors, designed by Siegmund Riefler of Munich (1847-1912). But it was not until the 1920s that the first real breakthrough occurred – the Shortt free-pendulum clock, one of the most important developments in timekeeping since the invention of the pendulum clock itself two hundred years before. The free-pendulum idea had been pioneered by R. J. Rudd as long ago as 1899: a practical system was perfected in 1921-4 by William Hamilton Shortt, a railway engineer, working in conjunction with F. Hope-Jones and the Synchronome Co. Ltd. In ordinary pendulum clocks the free swinging of the pendulum, on which timekeeping accuracy depends, is interfered with by the need to sustain the pendulum's motion and to count the swings to tell the time. In a free-pendulum clock, these two functions are carried out by a subsidiary 'slave clock', allowing the master pendulum to swing

Opposite. The original mechanism and the dial of one of the two year-going clocks that Thomas Tompion (1639–1713), the 'father of English clockmaking', built for John Flamsteed. The pendulum, 13 ft long, and swinging above the clock dial, was later reduced in length and hung below the movement, to convert the clock to domestic use. The dial shown is one of the replicas in the Octagon Room at Greenwich (see p. 32). The original clock returned to Greenwich in 1994.

Above. This is Shortt's free-pendulum clock no. 16, master and slaves, at Greenwich about 1930, which controlled the Greenwich time signals from 1927 to 1940.

freely except for a fraction of a second each half-minute, when it receives an impulse from the slave. Before this, the best clocks had an accuracy of about 1 second in ten days: the Shortts were accurate to 10 seconds in a year. Greenwich acquired its first Shortt free-pendulum clock in 1924 when No. 3 took over as Sidereal Standard. Others followed. Soon, the free-pendulum clocks ousted all others for time-service purposes; some (by George Graham) had been in use for nearly 200 years, and none (except a copy of a Riefler) was less than 55 years old.

One consequence of the increased accuracy of the primary time-keepers was a change of concept in the operation of the Greenwich time service. When Airy instituted the service in 1852 he based the timekeeping on two standard clocks, the Sidereal Standard and the Mean Solar Standard. Radio time signals, however, made it possible to compare clocks in other observatories the world over with great precision several times daily. Furthermore, Greenwich itself had many more very accurate clocks. This led in 1938 to the abandonment of Airy's standard-clock concept in favour of taking the mean of several clocks, some keeping sidereal, some solar time, initially five at Greenwich and one at the National Physical Laboratory, Teddington, England (NPL), to which was added a year later one from Edinburgh. All were Shortt free-pendulum clocks.

Quartz crystal clocks

We must now consider briefly concepts of time, in particular the difference between an *instant* of time (the 'date', or the 'epoch'), and an *interval* of time. Someone catching a train or aircraft is interested primarily in the instant; a boxing referee for example, in the interval. There is a third concept and that is the *frequency* of a periodic phenomenon – the number of cycles of this phenomenon per unit of time: the name of the unit of frequency today is the hertz (Hz), identical with the older unit of a cycle per second (cps).

It was the search by telecommunications engineers for a reliable standard of frequency of electro-magnetic waves which gave rise to the development of the quartz crystal clock, destined to prove an even more significant innovation than the free-pendulum clock had been some ten years earlier. The quartz crystal was developed with the advent of radio broadcasting in the early 1920s, giving for the first time a highly stable radio-frequency source. The first quartz clock proper was described by Horton and Marrison of the USA in 1928. The first quartz clock at Greenwich was installed in 1939, and was of a type developed by Essen at the NPL, with an accuracy of about 2 milliseconds per day. (A millisecond (ms) is one thousandth of a second; a microsecond (μs) one millionth of a second.) There were plans to install further quartz clocks at Greenwich but these were postponed by the outbreak of war and the transfer of the time service to the magnetic observatory at Abinger in Surrey, which was thought to be safer from enemy action. A reserve time-service station was set up at the Royal Observatory in Edinburgh in 1941. Although no quartz clocks were available at Abinger initially, information from two at the NPL was transmitted daily and incorporated, with the free-pendulum clocks, into the 'mean clock'.

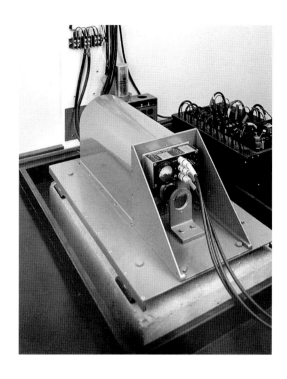

The last quartz crystal clock in use at the RGO, in 1978, at Herstmonceux, Sussex. Made in the USA, it was regulated by a lenticular AT-cut crystal oscillating 2.5 MHz.

The needs of war required the British time service to achieve a tenfold increase in the accuracy of radio time signals, in connection with the development of radar, and for precision air navigation systems. Arrangements were therefore made with the Post Office Radio Branch in 1942 for information from their own quartz clocks to be transmitted daily to Abinger. Their performance was so good that in 1945 the Shortt clocks ceased to be part of the 'mean clock'. The quartz clocks, their errors determined by astronomical observations at Abinger and Edinburgh, became the primary standards on which the time service was based, while the observatory clocks became secondary standards used for the control of time signals. Then, in 1944, the control of the international time signal from Rugby was taken over by new quartz clocks at Abinger, as was, in 1949, the control of the BBC's six-pip signal. The time-service station at Edinburgh closed down in January 1946 and, shortly after, six quartz clocks were installed at Greenwich, though the headquarters of the time service remained at Abinger which had twelve quartz clocks. Accuracies had meanwhile improved to about 0.1 milliseconds a day. Meanwhile the astronomers had moved away from the smoke and street lights of Greenwich the clearer air of Herstmonceux in Sussex, where the observatory became known as the Royal Greenwich Observatory (RGO). The time service moved from Abinger to Herstmonceux in 1957.[1]

Herstmonceux castle, Sussex, home of the Royal Greenwich Observatory from 1957 to 1990.

The Isaac Newton 2.5 m reflecting telescope at Herstmonceux, before it was relocated to La Palma in the Canary Islands.

Far left. In the early 1950s, when quartz technology in the Soviet Union was undeveloped, Feodosii M. Fedchenko of VNIFTRI (the Russian national physical laboratory) designed what is claimed to be the world's most accurate pendulum clock as an alternative, based on a pendulum swinging in a vacuum cylinder. About 45 were made for observatories and television centres, and this example presented by VNIFTRI to the Old Royal Observatory at Greenwich in 1994, was the first to leave the old Soviet Union, apart from one in Cuba.
Left. The first commercially produced radio-controlled wristwatch. In Britain, this picks up the radio time signal from the National Physical Laboratory transmitter near Rugby. The aerial is in the watch strap.

The non-uniform Earth

All these improvements in precision drew attention to another problem, ably summed up by Sir Harold Spencer Jones, tenth Astronomer Royal, in 1950:

> The rotation of the Earth provides us with our fundamental unit of time – the day. The first requirement of a fundamental unit is that it should be constant and reproducible, the unit should mean the same thing to all men and at all times. In taking the day, or, more precisely, the mean solar day as the fundamental unit from which we derive the hour, the minute, and the second as subsidiary units, it has been implicitly assumed that its length is invariable or, in other words, that the Earth is a perfect time-keeper.[2]

That the Earth was *not* a perfect timekeeper had been postulated by Immanuel Kant as long ago as 1754, but to get the full story we must go back another sixty years. In 1695 Edmond Halley, through an analysis of ancient eclipses, had come to the conclusion that the Moon's motion round the Earth was accelerating, and this was later confirmed by direct measurement. In 1787 Laplace showed that this could be explained by slow changes in the shape of the Earth's orbit, but in 1853 Adams pointed out that these orbital changes would account for only half the observed value of the Moon's acceleration. After much debate, it was eventually shown that that part

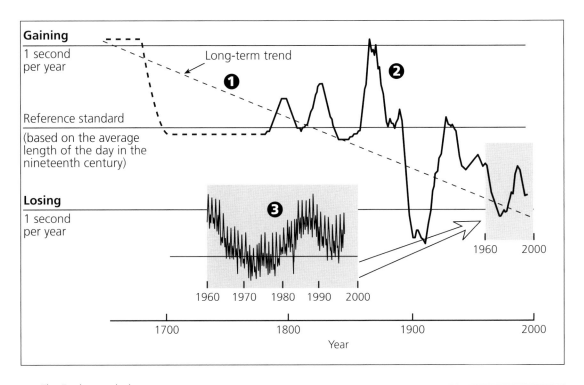

Gaining
1 second per year

Long-term trend

Reference standard
(based on the average length of the day in the nineteenth century)

Losing
1 second per year

1700   1800   1900   2000

Year

The Earth as a clock, showing changes in the Earth's rotation rate over the last 300 years. The graph shows the three different aspects of this referred to on p. 169: (1) long-term trend (or secular changes); (2) irregular (and unpredictable) fluctuations; (3) annual seasonal fluctuations.

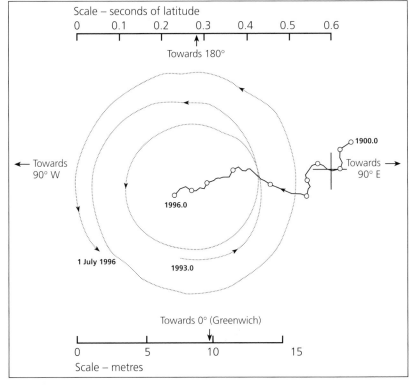

Scale – seconds of latitude

0   0.1   0.2   0.3   0.4   0.5   0.6

Towards 180°

← Towards 90° W

1900.0

Towards 90° E →

1996.0

1 July 1996

1993.0

Towards 0° (Greenwich)

0   5   10   15

Scale – metres

Motion of the mean pole, 1900-96 (solid line), with detailed motion from 1993 (pecked line). From *International Earth Rotation Service 1995 Annual Report.*

of the Moon's acceleration not accounted for by Laplace's gravitational theory could be explained by assuming a progressive slowing down in the rate of the Earth's rotation, due largely to the friction caused by the tides.

We now know that changes in the rate of rotation of the Earth occur on broadly three different time scales:

(a) *secular changes:* the progressive slowing down already referred to, caused by tidal friction and changes in the figure (the shape) of the Earth due to 'post glacial rebound' from the melting of the ice caps after the last Ice Age, amounting to an increase in the length of the day of 1.7 milliseconds per century;

(b) *irregular (and unpredictable) fluctuations,* probably caused by differing rates of rotation between the molten core and the solid mantle of the Earth, which may result in the length of the day increasing or decreasing by up to 4 milliseconds in a decade;

(c) *seasonal variations,* reflecting the changing patterns of winds through the season; for example, the SW Monsoon in Asia and the Indian Ocean the nothern summer, and the NE Monsoon in the winter. Momentum is exchanged between the atmosphere and the mantle/crust, mainly at the large mountain ranges.

There is another phenomenon which, though it does not affect the Earth's speed of rotation, has to be taken into account when great timekeeping precision is needed. This is 'polar variation', or the wobble of the Earth on its axis (rather like the effect of a 'sloppy' bearing in a machine). This causes the Earth's poles to wander in an approximately annual circular path with a radius of about 8 metres. The effect of polar variation is to change the geographical latitude and longitude of every place on Earth (when ascertained by astronomical means) by a minute amount, the consequence of which is that there are minute variations in the time scale at each place due to changes of longitude.

As Spencer Jones said, the first requirement for a fundamental unit is that it should be constant and reproducible. By 1950 the second of time based upon the rotating Earth had been proved to have variations in length which, although insignificant hitherto, could no longer be neglected. What was to be done?

Ephemeris time

The first solution was to abandon the solar day as the fundamental unit and to substitute the year, the length of which, though not constant, can be predicted, decreasing by about 0.5 seconds per century. This led to the introduction internationally in 1952 of a new kind of time scale for certain purposes – Ephemeris Time (ET), which is used (as the name implies) in compiling the various national ephemerides and almanacs. But first we must say something about

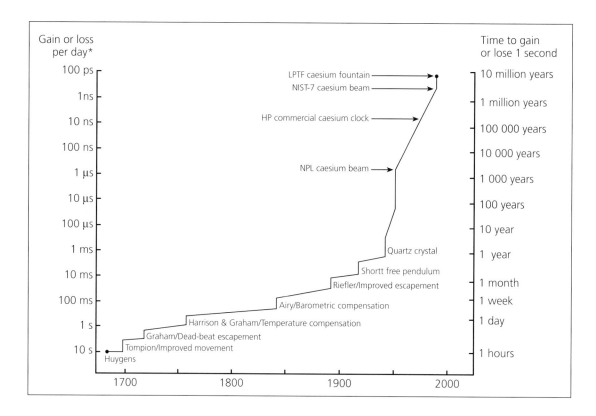

Gain or loss per day*

100 ps	
1 ns	
10 ns	
100 ns	
1 μs	
10 μs	
100 μs	
1 ms	
10 ms	
100 ms	
1 s	
10 s	

Time to gain or lose 1 second

10 million years
1 million years
100 000 years
10 000 years
1 000 years
100 years
10 year
1 year
1 month
1 week
1 day
1 hours

LPTF caesium fountain
NIST-7 caesium beam
HP commercial caesium clock
NPL caesium beam
Quartz crystal
Shortt free pendulum
Riefler/Improved escapement
Airy/Barometric compensation
Harrison & Graham/Temperature compensation
Graham/Dead-beat escapement
Tompion/Improved movement
Huygens

1700   1800   1900   2000

The increasing accuracy of precision clocks (order of magnitude only).
• NPL (National Physical Laboratory, Teddington);
• HP (Hewlett Packard) model 5071A caesium clock (commercial);
• Clock NIST-7 (US National Institute of Standards and Technology), an optically pumped caesium-beam standard with uncertainty of less than 0.9 ns/day;
• Clock LPTF-F01 (Laboratoire Primaire du Temps et des Fréquences, Observatoire de Paris), a caesium fountain frequency standard with uncertainty of about 0.3 ns/day.

Universal Time. As we saw in the last chapter, Greenwich Mean Time came to be known by astronomers and other scientific users as Universal Time (UT), as a result of the decisions of the Washington conference in 1884 and the specific recommendations of the IAU in 1928: so in this chapter, we shall speak of UT rather than GMT when referring to the mean solar time on the Greenwich meridian. Now, UT is based on the spin of the Earth on its axis and is the time-scale needed for celestial navigation. But as we have seen, the rate of spin is variable, so in 1956 it became necessary to define UT more closely for specialist use in the time services:

UT0 Mean solar time of the Prime Meridian obtained from direct astronomical observation.

UT1, UT0 corrected for the observed effects of polar motion (maximum about 0.035 seconds). UT1 is the time-scale used in celestial navigation.

UT2, UT0 corrected for observed polar motion *and* for extrapolated variations in the Earth's rotation rate (also up to about 0.035 seconds). UT2 is a 'smoothed' time-scale giving as uniform a time as possible, on which time signals before 1972 were based.[3]

The full story of ET and of its relationship to UT is too complex to tell here. Suffice it to say that ET was made to conform

* ms = millisecond .001 s; μs = microsecond .000 001 s; ns = nanosecond .000 000 001 s; ps = picosecond .000 000 000 001 s.

closely to UT by making the Ephemeris day the average length of the mean solar day during the nineteenth century. Finally, in 1956, the mean solar day was abandoned internationally as the fundamental unit of time in favour of the Ephemeris Second, defined as 'the fraction 1/31 556925.9747 of the tropical year for 1900 January 0^d 12^h ephemeris time'.[4]

This change, however, did not solve all problems. Because it was invariable, a second defined in terms of Ephemeris Time suited many theoretical needs and was immediately adopted for the various ephemerides. But it was not suitable for everyday use for two main reasons: first, it was not readily accessible and could only be determined to the required accuracy after a long delay, several years' worth of observations having to be analysed before any kind of result could be obtained; secondly, those who were interested in the precise instant of time as opposed to the time interval – and these included the public at large – required that time signals should conform fairly closely to the rotation of the Earth, to the alternation of night and day. Though the differences between ET and UT were very small in any one year, the accumulated differences could become significant because of the systematic slowing down of the Earth's rotation. By 1952 when ET was first brought into use, there was an accumulated difference of almost 30 seconds between ET, based on the nineteenth-century rotation rate, and UT, based on 1952.

The use of Ephemeris Time in time signals proved to be a compromise because the physicist and telecommunications engineer, who required that the length of the second on their time signal should be uniform – that it should 'mean the same thing to all men and at all times' – while the ordinary user (and the navigators and surveyors) required that when a time signal said it was noon, then it should be noon according to the heavens. Before 1944 time signals controlled by Greenwich had been kept as closely as possible in line with the Earth's rotation, resulting in a second (as obtained from time signals) that could vary in length from day to day, albeit very minutely. Then, from 1944, an effort was made to transmit time signals at as uniform a rate as possible, based on the mean of the best quartz clocks available, with 'jump' corrections introduced when necessary (on Wednesdays) to maintain agreement with universal (astronomical) time. In the USA at that date, however, no such attempt at a compromise between time and frequency was made: time signals from radio station NSS (Annapolis), controlled by the US Naval Observatory, were kept strictly in line with the Earth's rotation while standard frequency transmissions from radio station WWV controlled by the US National Bureau of Standards were kept at as uniform a rate as possible.

The atomic clock

The answer to the first of the disadvantages of Ephemeris Time, inaccessibility, proved to be the atomic clock. The first operational complete atomic clock system was developed at the US National Bureau of Standards, Washington, DC, by Harold Lyons and his associates in 1948–9, using an absorption line of ammonia to stabilize a quartz crystal oscillator. First brought into operation on 12 August 1948, it was developed specifically for use as a frequency standard. However, attention soon turned away from ammonia towards the element caesium. Most of the early development of the caesium standard, with which the names of J. E. Sherwood, J. R. Zacharias and N. Ramsay are particularly associated, took place in the USA. But it was at the National Physical Laboratory, England, that a caesium beam standard was first used on a regular basis. Designed by L. Essen and J. Parry, this was brought into use for the calibration of quartz clocks and frequency standards in June 1955, the very year that the decision was taken that the ephemeris second, with all its limitations, should be adopted as the fundamental unit of time. During the next few years further laboratory-type caesium standards were brought into use in Boulder (Colorado), in Ottawa and in Neuchâtel.[5]

Even these earliest examples of atomic clocks were a hundred times more stable in the long term than the best quartz crystal standard and did not suffer from the drifts in rate associated with the 'ageing' of quartz crystals. Subsequent developments from them have provided a time-scale of great uniformity, of very high precision (at least ten times more accurate than the prototype), and almost immediate accessibility. But it was some years before all this could be realized. The latest super-tube atomic caesium standards achieve a short-term stability comparable with that of a free-running quartz clock.

All clocks have to be adjusted so that they go at the correct rate – that they 'keep time' – and also have to be set to time. The new atomic clocks were no exception and the first task was to calibrate them against accepted standards; in other words, the atomic time-scale had to be related to the astronomical time-scale. Between 1955 and 1958 atomic clocks in the UK and USA were calibrated against astronomical time-scales at Herstmonceux and Washington. The first atomic time-scale was known as GA (Greenwich atomic), using initially the NPL caesium standard with a rate set to agree as closely as possible with Ephemeris Time. Then in 1959 the US Naval Observatory's time-scale A1 was brought into general use. Its point of origin was arbitrarily set so that Atomic Time and UT2 were the same at midnight on 1 January 1958. At the same time the atomic second was defined in terms of the resonance of the caesium

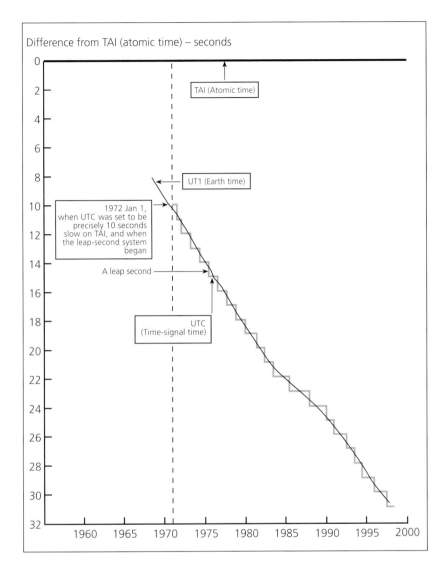

Difference from TAI (atomic time) – seconds

TAI (Atomic time)

UT1 (Earth time)

1972 Jan 1, when UTC was set to be precisely 10 seconds slow on TAI, and when the leap-second system began

A leap second

UTC (Time-signal time)

Time signals and the leap second. Since 1972 TAI (atomic time) and UTC (time-signal time) run at the same rate. Because they were based on the nineteenth-century rate of the Earth's rotation, they are faster by almost one second a year than UT1, which is derived from today's rotation rate. UTC (time-signal time) is kept within 0.9 s. of UT1 (Earth time) by delaying UTC by a leap second at the end of June or December (the downward steps) so that UTC falls behind UT1 and then starts to catch up again. Each leap second makes UTC one more second behind TAI. There have been twenty-one leap seconds since 1972.

atom. In 1964 the atomic second was recognized internationally as a means of realizing the ephemeris second. Then at the 13th General Conference of Weights and Measures, Paris, 1967, the astronomical definition of a second was abandoned in favour of the atomic second as the fundamental unit of time in the International System of Scientific Units (SI system). This new definition was set out in the following resolution:

> That the unit of time in the International System of Units shall be the second, defined as follows:- The second is the duration of 9 192 631 770 periods of the radiation corresponding to the transition between the two hyperfine levels of the ground state of the caesium-133 atom.[6]

As other caesium clocks came into use all over the world – clocks which could be compared with each other by radio and other means to an accuracy of a microsecond or better – it became possible to form an international 'mean clock' of great precision, the larger number of independent contributions achieving very great uniformity. These clocks agreed with each other to within a few microseconds in a year, whereas they differed from the time-scale of the rotating Earth (Universal Time) by up to a second in a year.

The Bureau International de l'Heure which had co-ordinated international timekeeping since 1919, had formed its own atomic time-scale (A3) soon after that of the USA, based on the results from three independent atomic time scales, each comprising a certain number of caesium standards in England, Switzerland and the USA, with its origin on 1 January 1958. The BIH atomic time-scale was formally adopted internationally in 1971, the name being changed to International Atomic Time (TAI). It is worth noting that, on 1 January 1997, thirty-nine years after the two scales had arbitrarily been set together, UT1 (based on the Earth's actual rotation rate 1958-96) had lost some thirty seconds on TAI (based on the Earth's rotation rate in the nineteenth century).

Co-ordination of time signals

To return to time signals, in 1958 the UK time service adopted a system later called Co-ordinated Universal Time (UTC), the aim of which was to keep the signals within approximately a tenth of a second of UT2. This was achieved by slightly changing the rate of the atomic clocks generating the time signals, applying an 'offset' which caused them to approximate to the current rate of UT2. (In the

Hewlett-Packard 5071A caesium clock. The Royal Greenwich Observatory ceased to originate the Greenwich time signal, using Hewlett Packard atomic clocks, in 1990. This commercially produced model, introduced in 1992, now accounts for over half the time standards incorporated in the world's UTC time scale, co-ordinated by the Bureau International des Poids et Mesures. It also provides the main time reference for the global satellite navigation system (GPS). These clocks are regulated by the decay of radioactive caesium.

1960s this was a losing rate in respect of Atomic Time.) The offset value remained unchanged through each calendar year but, to cater for unforeseen changes in the Earth's rotation rate, step corrections were applied each month to keep UTC within 0.1 seconds of UT2. In 1961 full co-ordination between UK and US time services was instituted. The time signals were synchronized, and the same annual offsets and monthly step corrections applied. In 1963 this UK/US system was extended worldwide, being operated under the control of BIH in Paris. It was then that the name Co-ordinated Universal Time was adopted.[7]

However, the increase in number and sophistication of satellite and other electronic communication and navigation systems brought to light major practical difficulties. These systems depend upon the very precise synchronization of both the radio signals themselves and of frequencies. The resetting in time in line with the step corrections proved difficult and inconvenient; the resetting in frequency once a year even more so. The fact that the second as broadcast in time signals did not conform to the legal second was an aesthetic rather than a real difficulty.

The leap second

After much discussion at all levels, national and international, fundamental changes were made in the system of time signals worldwide. As from 1 January 1972 time signals radiated atomic seconds precisely, the new UTC being set to be exactly 10 seconds slow on TAI as from that date. This is the system in operation at the time of writing.

It was originally agreed that time signals should not be allowed to depart more than 0.7 seconds (a tolerance later increased to 0.9 seconds) from UT1, the time-scale needed by navigators and astronomers. To achieve this, step corrections, either positive or negative, of exactly one second, known as 'leap seconds', are made when needed on the last day of a calendar month, preferably 31 December and/or 30 June. Just as the fact that there is not a whole number of days in a year gave rise to the leap year, so the fact that there is not a whole number of atomic seconds in a solar day gave rise to the leap second.

Thus international time and frequency radio signals, such as those from station

A hand-held GPS set in use. Its position, shown as North (N) latitude and longitude on the second and third lines of the screen, is fixed by the simultaneous reception of signals from several satellites in 12-hour orbits. The instrument is is being held inches to the west (W) of the Longitude zero line at the Old Royal Observatory and is registering the fact.

MSF in the UK or WWV in the US, radiate exact atomic seconds without interruption or variation throughout the year. All that happens when a leap second is added or subtracted is that the numbering of the seconds markers is changed. Thus, if a positive leap second is needed on 31 December because UTC is seen to be drifting too far from UT1, the last 'minute' of the year will have sixty-one seconds instead of the conventional sixty. For a negative leap second, the last 'minute' would have only fifty-nine seconds.

To cater for those who must know UT1 more precisely (navigators and astronomers for example), the principal time and frequency signals contain a form of code which indicates the number of tenths of a second that the time signal (UTC) on that particular day is fast or slow on astronomical time (UT1).

Up to 1987, the world's time signals were coordinated by the BIH in Paris, based on a world 'mean clock' comprising some eighty atomic clocks from twenty-four participating countries. (Only those countries within the coverage of the radio navigational aid Loran-C were at that time able to participate.) It was the BIH who decided when leap seconds should be inserted. In 1972, UTC was set to be precisely 10 seconds slow on TAI: since then, twenty positive leap seconds have had to be inserted. On 1 January 1997, therefore, UTC lagged exactly 30 seconds behind TAI, which always differs from UTC by an exact number of seconds.

On 1 January 1988, the BIH ceased to operate, its rôle as coordinator of atomic time being taken over by the Bureau International des Poids et Mesures (BIPM) in Sèvres, outside Paris, a body well-established under international treaty, the Convention du Mètre. The duty of reconciling atomic time with astronomical observations (the decision on when to insert leap seconds and the publication of differences between UT1 and UTC, information on polar motion, etc.) now rests with the International Earth Rotation Service (IERS) whose central office is at Paris Observatory.

The world mean clock is currently based on around 220 atomic clocks in 38 laboratories worldwide. Loran-C is no longer used for comparison of time scales between these laboratories but use is made of the Global Positioning System (GPS) which relies upon precise time and position information (from about 24 satellites in orbit, carrying atomic clocks) to allow anyone anywhere with an unobstructed view of the sky to find both the time (UTC) and their position with an accuracy of around a ten-millionth of a second and 30 metres, within at most a few minutes, using equipment costing under £200. (Russia operates a similar system known as GLONASS.)

For comparison of time between laboratories, a system known as 'common-view GPS' is used. The laboratories program special GPS timing receivers to receive from particular satellites in turn for 13-minute periods so that all laboratories within view of the satellite are receiving it at exactly the same time. This 'common view' eliminates many of the sources of error and allows time differences between laboratories to be measured to within 10 nanoseconds. An even more precise method, offering accuracies better than 1 nanosecond, is now being put into use between some laboratories. This entails simultaneous transmission of timing signals both ways using a transponder of a communications satellite in geostationary orbit.

Turning to national affairs in Britain, the Greenwich Time Service provided time for the kingdom from the Royal Observatory Greenwich, from 1852; from Abinger in Surrey, from 1940, during the Second World War; and then, after 1957, following the Observatory's move from Greenwich, from the newly named Royal Greenwich Observatory at Herstmonceux in Sussex. As well as the Greenwich time-ball, the principal means of time dissemination was first the once-daily electric telegraph, then, from 5 February 1924, the six-pip time signal every 15 minutes by landline to the BBC in London, which broadcast them as necessary.

At Herstmonceux, the role of the Royal Greenwich Observatory (RGO) changed progressively from what it had been at Greenwich, and the measurement of time by electronic or astronomical means became less prominent in its work. It turned to astrophysics, building at Herstmonceux the 2.5m Isaac Newton Telescope (INT) for astronomical observations of the nature of stars and galaxies. This was moved to the higher, dryer, clearer mountain site of La Palma in the Canary Islands in 1984, where the RGO built an even bigger instrument, the 4.2m William Herschel telescope, as well as the 1m Jacobus Kapteyn Telescope. The RGO developed into a centre of technology supporting the new island observatory, maintaining the telescopes and their instruments, ensuring that they reached and remained at the forefront of astronomical research. It was very successful in developing the island site, which gradually became more self-sufficient in its maintenance needs and in transferring its instrument-building skills into the larger community that grew up using the new telescopes. This new role demanded greater co-operation with the university community, so in 1990 the RGO moved to Cambridge, adjacent to the old university Observatory, and close to the Hoyle Building of the Institute of Astronomy. At its new home, the RGO built instruments and other equipment for the new 8m

Humphry M. Smith, Head of the Time Department at the RGO from 1936 to 1976, Chairman of the Directing Board of the Bureau International de l'Heure, 1966–81, and head of various related commissions, was one of the most important international scientists concerned with GMT. He was a delegate, representing the UK, the BIH or the CCIR, at meetings of the international organizations concerned with time determination and dissemination, the introduction of atomic time and, principally, in the establishment and worldwide adoption of Coordinated Universal Time (UTC).

Gemini telescopes in Hawaii and Chile. Its history in this period was paralleled by that of the Royal Observatory in Edinburgh, which similarly built, developed and instrumented telescopes on Hawaii. As a result, the two observatories became competitors for the funds available for ground-based telescopes, which, after the Gemini instruments were completed in the late 1990s, decreased in favour of other kinds of astronomy.

The situation was becoming untenable and in July 1997 a decision was taken to merge the two organizations on a single site at Edinburgh. In making the announcement, the Minister of Science, John Battle, expressed a desire to see a new role for the RGO – Britain's oldest surviving scientific institution – possibly back at Greenwich. At the time of writing, this role has yet to be developed by the Ministry of Science, but it seems that history may eventually come full circle.

At the time of the RGO move to Cambridge, 5 February 1990, the Greenwich Time Service transmitted its last ever pips at one o'clock in the afternoon – 138 years after the service was inaugurated by George Airy, and 66 years to the day after the first six-pip time signal was broadcast.[8] Since 1990, the BBC has originated their own time signals, based on signals from the GPS satellite network and from the 60 kHz radio transmitter at Rugby (call sign MSF), which is operated by BT International under contract to (and monitored by) the National Physical Laboratory. Incidentally, a large proportion of the radio-controlled clocks in use in Britain today, both public private, depend upon time signals from MSF.

With the introduction into time signals in 1972 of the UTC tied to atomic time (TAI), in place of the old UTC tied to mean solar time (UT2, which non-scientists continued to call GMT), the arguments about the nomenclature of time-scales began again. The new time-scale continued to be based on an international meridian that approximates to the Greenwich meridian, but it could no longer be *defined* as mean solar time on the Greenwich meridian (GMT) even though it might never depart more than nine-tenths of a second from it.

Indeed, even the Greenwich meridian itself is not quite what it used to be – defined by 'the centre of the transit instrument at the Observatory at Greenwich'. Although that instrument still survives in working order, it is no longer in use and now the meridian of origin of the world's longitude and time is not strictly defined in material form but from a statistical solution resulting from observations of all time-determination stations which the BIPM takes into account when co-ordinating the world's time signals. Nevertheless, the line in the old observatory's courtyard today differs no more than a few metres from that imaginary line which is now the Prime Meridian of the world.

Though no longer in use in astronomy proper, the term GMT continued to be used by navigators, for many civil uses, and as a description of statutory legal time in many countries. However, there was opposition to the continuance of this state of affairs, particularly from France. In 1975 the 15th General Conference of Weights and Measures (CGPM) recommended to its adhering countries that, as the new UTC was used in time signals, it should in future be used as the basis of legal time:[9] it would thus replace GMT which, because of the change in derivation of UTC in 1972, was now ambiguous.[10] France issued a decree on 9 August 1978, repealing the law of 1911 (which said that legal time in France was Paris mean time retarded by 9 minutes 21 seconds) and ordaining that legal time in all French territories should in future be obtained by adding or subtracting an exact number of hours to UTC, which can be increased or reduced during part of the year (thus allowing for Summer Time), and that the term GMT was not to be used in any future regulations.[11] Almost all major countries have now adopted UTC as the basis for legal time, although the UK has yet to take this formal step. Nevertheless, it is interesting that 1994 European Union legislation on harmonizing the instants of change to and from daylight saving time[12] is written in terms of Greenwich Mean Time.

Because one extra atomic second has to be added some years, it might be thought that these years are actually longer than the others. That is not so. So far as is known, the length of a year is decreasing by only some half a second per century. Thus, 365 2000-type days are longer by about a second than 365 nineteenth-century-type days on which the time signals are based. Atomic time (TAI) currently runs faster than astronomical time (UT) because its seconds are shorter, having been based on the nineteenth-century year. Gradually, but erratically, the noon 'pip' time signal gets ahead of noon according to the stars. It will not be allowed to get more than 0.9 seconds ahead; before then a positive leap second will have been introduced to move it back by exactly one second to start catching up again.

Leap seconds are introduced either at the end of the UTC year or at the end of UTC June – at the same instant worldwide. The UTC time goes 23:59:58, 23:59:59, 23:59:60, 00:00:00, 00:00:01, the last minute of the UTC day having 61 seconds. In principle, a negative leap second can be introduced (omitting Second 59) if ever a correction is needed because atomic time is falling behind, but that is most unlikely to happen in the foreseeable future.

It is not possible to predict how Earth's rotation rate will change over the next few decades. At present, it is slowing down considerably faster than the average over the last three hundred years. However, it could be that the trend will be reversed and that, should the present system of time signals persist, no leap second, or even negative leap seconds, will have to be inserted by, say, the 2010s. Nevertheless, it is inevitable that, some time in the future – perhaps in tens of years, perhaps in hundreds, perhaps even in thousands – there will be a need for two and then three positive leap seconds each year, should we continue to base our time-scale on the average length of the day in the nineteenth century. As for the more distant future, one of the effects of the decelerating Earth will be that the need for leap years (but not for leap seconds) will disappear: in a few million years' time, there will only be 365 days in the year, not 365¼ as now.

Conclusion

In this book we have told how, when the Royal Observatory was founded three hundred years ago, Greenwich time concerned only one astronomer and his assistant; how in the 1760s the publication of *The Nautical Almanac* and the development of the marine chronometer meant that the Greenwich meridian and Greenwich time began to be used by mariners of all nations; how, ashore, at the turn of the nineteenth century, the increasing use of clocks and watches led to the abandonment of apparent, or 'sundial', time and the adoption of mean, or 'clock', time as the local time kept in each community; how, not many years later, the development of railways led to the abandonment of local time and the adoption in each country of a national, or 'railway', time; how, by 1884, the development of worldwide communications was such that there was a need for international, or 'universal', time, and how it was Greenwich time that was chosen for this both ashore and afloat; how in the 1940s and 1950s the development of quartz and atomic clocks (the latter more accurate than Flamsteed's pendulum clocks by a factor of some 8 million) resulted in the adoption of atomic time and the abandonment of the Earth as the fundamental timekeeper, the Earth's rotation rate having proved, after all, not to be uniform.

Britain's time signals no longer emanate from Greenwich itself. The world's time is now co-ordinated from Paris (and is called Universal Time). Universal Time today truly is universal, being based on the mean of clocks from twenty-five different nations. Nevertheless the world's Prime Meridian for longitude and time still passes through the old observatory at Greenwich. Although the world's time signals (and in some countries legal time) are no longer based on GMT as strictly defined, nevertheless the time-scale on which they are based – UTC – is bound to be within a second of the old GMT.

And navigation, now so much transformed by advances in electronics and space science, is still dependent on the measurement of time, to a precision undreamt of thirty years ago.

Appendices

I FINDING THE LONGITUDE

The difference in longitude between any two places on the surface of the Earth is precisely equal to the difference between the local times of the two places. It is this unalterable relationship which has made it necessary for a book such as this, primarily about time, to have a sub-title 'and the longitude'. Furthermore, until recent times, many of the most important developments in timekeeping have been directly stimulated by the need to find longitude at sea. The purpose of this appendix is to explain in simple terms why this should have been so: it should be read in conjunction with Chapter One which gives the historical facts.

As explained in that chapter, geographical positions have been described in terms of latitude and longitude at least since the time of Ptolemy in the second century AD. Finding latitude has been possible since ancient times, by making some sort of measurement of the Sun's altitude at noon, at first by measuring the lengths of shadows of a gnomon of known height, then by more sophisticated instruments developed for the purpose, initially ashore and then, from the fifteenth century, at sea too. Latitude could be found also by measuring the height of the Pole Star above the horizon, although an allowance had to be made for the fact that the star is not situated exactly at the celestial north pole (which in northern latitudes is always elevated above the horizon at precisely the amount of the observer's latitude).

A practical method of finding longitude, however, did not come within reach until the last few hundred years. Nevertheless, the theoretical solution is comparatively simple (see diagram, below) and was known to Hipparchos, for example. When the Sun is on your meridian and it is 12 noon to you at *G*, it is 6 a.m. (and

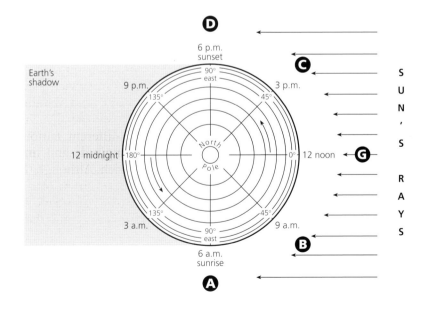

Longitude and time (the Earth shown as at the equinox).

at the equinox it would be sunrise) for person *A* situated on a meridian 90° *west* of yours, while it would be 9 a.m. for person *B* 45° to the west. Similarly it would be 3 p.m. for person *C* 45° *east* and 6 p.m. for person *D* 90° *east*. An hour later at 1 p.m. for *G*, it would be 7 a.m. for *A*, 10 a.m. for *B*, 4 p.m. for *C* and 7 p.m. for *D*: though the times have changed, the differences of times remain constant and are a measure of the differences of longitude between the various places. Hipparchos thought probably of difference of longitude in terms of time, of so many hours east or west: today, we usually think of it in terms of arc, of so many degrees, minutes and seconds east or west; but in this context, time and arc are completely interchangeable, 24 hours answering to 360°, 12 hours to 180°, 1 hour to 15°, and 4 minutes to 1°. So, if you know that, when it is 9 a.m. to you, it is at that moment 7 a.m. for person *E*, then the difference of longitude must be 2 hours, which is equivalent to 30°; because it is earlier in the morning for *E* than for *G* (and as the Sun appears to move from east to west), then *E* must be 30° *west* of *G*.

But how does person *G* know what time it is at, say, *E*, over a thousand miles to the west? One way of finding out would be to transport some form of timekeeper between the two places, but this did not become technically possible until the eighteenth century. Another method would be to record the time of some phenomenon which could be seen at both places simultaneously; then, at some later date, the time recorded at the two places could be compared and the difference of longitude found. A refinement of that method would be to predict the time at which the phenomenon would occur at some chosen place (say Greenwich) and then the longitude difference could be found immediately from the local time of the observed phenomenon without having to wait to compare actual measured times.

Longitude by lunar eclipse

Hipparchos is said to have proposed the use of a lunar eclipse as the phenomenon to be observed when Sun, Earth, and Moon are directly in line and the Earth's shadow moves across the Moon's surface, reaching any particular point at precisely the same instant for an observer wherever he may be. Ptolemy recommended this method of finding longitude ashore (though he gives only one example of its having been used) but he does not tell us how the local time of the phenomenon should be found. For the shadow to be seen on Earth at all, the Sun *must* be below the horizon, so a sundial could not have been used directly, and only rudimentary timekeepers would have been available – perhaps a water-clock, or hourglass.

Lunar-distance changes, relative to the Sun (as illustrated here) or to a zodiacal star.

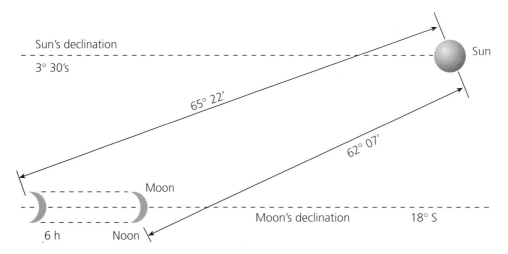

Sun's declination
3° 30's

Sun

65° 22'

62° 07'

Moon

Moon's declination   18° S

6 h   Noon

This is still something of a mystery. One could use the position of the stars, either by measuring the position angle of the Great or Little Bear in relation to the Pole Star (as was to be done with the nocturnal, an instrument invented over a thousand years later), or by measuring the altitude and finding the time by some form of planispheric astrolabe. Or was the Moon's shadow used in some way, remembering that it would be Full Moon so that, in the middle of the eclipse, the Moon would be exactly opposite the Sun? A Spanish method of 1582 for use in the Indies recommended the setting up ashore of a vertical gnomon precisely one-third of a yard long: the length of the *Sun*'s shadow when at its shortest the day before (or day after) the eclipse gave the latitude; the direction and length of the *Moon*'s shadow at the beginning or end of the eclipse could (after considerable calculation, generally carried out back in Spain) give the longitude. There are few reports of this elegant method having been used but, from an observation of an eclipse of 1584, the position of the Casa Real in Mexico City was calculated as being only 21 km too far south and 23 km too far west of the true position, a remarkable result even taking into account a large element of luck.[1]

Longitude by lunar distance

There were no fundamental developments in the longitude story until 1514, when Johann Werner described the lunar-distance method, which was the first to give some hope of finding longitude *at sea*. During the course of a month, the Moon appears to move one complete revolution from west to east against the background of stars. This comparatively fast movement (see diagram, above) – approximately ½°, or her own diameter, in an hour – prompted Werner to suggest using the Moon as a gigantic celestial clock, with the Moon as the hand and the stars in the zodiac as the figures on the dial. As seen from an observer on Earth on any particular night, the stars in the zodiac move across the sky from east to west at a fraction more than 15° per hour, while in the same period the Moon moves about 14½°. Thus the Moon appears to lag behind the stars at the rate of about half a degree per hour: a zodiacal star which is, say, 40° ahead of (that is, to the west of) the Moon at 22.00 will be 41° ahead at midnight (in fact, the star will have moved about 30° while the Moon moves 29°). Werner assumed that the 'lunar distance' between zodiacal star and the Moon at a particular instant would be the same whatever the observer's geographical position, and he therefore proposed that the changing lunar distance could be used as a measure of the time on some reference meridian (which we will henceforth call Greenwich). This could then be used to find longitude, provided the Moon's position with reference to the stars could be predicted in advance, and the actual distances in the sky measured to the required degree of accuracy.

Werner's assumption that the Moon's position in the sky is not affected by the observer's geographical position is fallacious. Making allowance for this – parallax – and for the·different amounts of refraction affecting the two bodies if they are at different altitudes, was to prove one of the most tedious pieces of arithmetic which the navigator eventually had to perform. Werner's basic method was, however, perfectly sound in theory, although there were three fundamental needs to make it work in practice: knowing the precise positions of the stars relative to each other; being able to predict several years ahead the Moon's position against that frame of reference (the navigator had to have these predictions before he sailed, and voyages could last several years); and having an instrument which could take the necessary observations to the required degree of accuracy. It was 250 years before these needs were satisfied.

Lunar distances became a practical possibility in the 1760s with the publication of the *Nautical Almanac* where the first two requirements were met by the tables of distances from the Moon to the Sun and certain zodiacal stars, predicted in the almanac for every three hours throughout the year. The third requirement was met by the sextant. Briefly, a lunar-distance sight required three simultaneous or near-simultaneous observations (see diagram, below) – the angular distance from the Moon to a star or to the Sun, the altitude of the Moon, and the altitude of the star or Sun, the times being taken by the best watch available. Though there were advantages in taking observations when the horizon was fully visible (in daylight or at twilight), this was not essential because great precision was not absolutely necessary for the altitude observations; so long as some horizon could be seen, this would suffice. The observations would then be 'reduced' as follows:

Step A – finding the local time of observation: from an altitude observation of Sun or star, preferably that taken at the same time as the lunar distance, but, if there had been a bad horizon, it could have been from an altitude taken the previous afternoon or the next morning, carried forwards or backwards by the watch: the local time was found by relatively simple spherical trigonometry;

Step B – clearing the observed lunar distance from the effects of refraction and parallax: tables were provided for doing this, but the calculations were laborious; the result was the correct lunar distance, at the time of observation, between the centre of the Moon and the centre of the Sun or star, as seen from the centre of the Earth – which is the form in which lunar distances are tabulated in the almanac;

Step C – finding the Greenwich time of observation: by entering the corrected lunar distance obtained in step B in the lunar distance table in the almanac, interpolating as necessary. For example, about 04.30 astronomical time (16.30 civil time) on 4 October 1772, an unknown navigator in an East Indiaman in the South Atlantic while on passage from Plymouth to the Cape of Good Hope, took four consecutive lunar-distance observations of the 'Limbs' (edges) of the Sun and Moon, the mean of which was 102° 26′ 55″. The corrected lunar distance (Step B) came to

Lunar-distance observations, showing the three near-simultaneous observations required to find longitude by lunar distance: (1) the angular distance between the Moon and selected star (or Sun); (2) altitude of Moon above Horizon; (3) altitude of star or Sun above horizon.

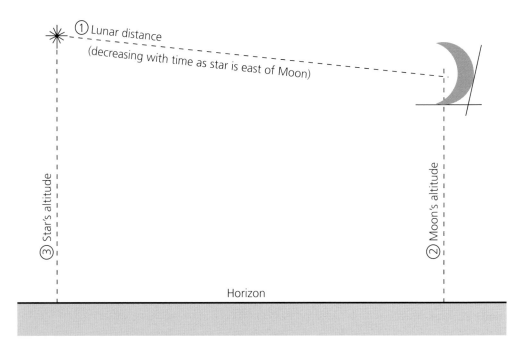

102° 36′ 08″. Then, interpolating in the lunar distance table in the *Nautical Almanac* between the value given for 6 h and that for 9 h (for the Sun on 4 October on p.65 above), the Greenwich time of the observation proves to have been 6 h 23 m 39 s.[2]

Step D – finding the longitude: by taking the difference between the local time found in Step A and the Greenwich time found in Step C. Thus, taking the same example:

	h	m	s
Local time from Step A	04	26	56
Greenwich time from Step C	06	23	39
Longitude in time	01	56	43
= Longitude in arc	29° 11′ West		

(To find whether east or west, repeat the jingle 'Longitude east, Greenwich time least; longitude west, Greenwich time best.' The example is to full numerical accuracy, but an error of 10″ in the observed distance corresponds to a minimum error of about 5′ in longitude.)

Longitude by eclipses of Jupiter's satellites

The invention of the telescope in the first few years of the seventeenth century revealed the existence of four bright satellites revolving around Jupiter: like our own Moon, they were eclipsed periodically when they passed into the planet's shadow. Galileo pointed out that, if these could be predicted and the eclipses accurately timed, they could be used for finding longitude exactly as with the eclipses of our own Moon, but with the added advantage that they occurred much more frequently and the duration of observation was much shorter. This method was used extensively ashore but proved impracticable at sea because of the difficulty of observing. Tables giving predicted times are given in the almanacs.

Longitude by chronometer

As described in Chapter 3, this became a practical possibility in the second half of the eighteenth century. The basic principle for finding longitude was similar to that described for lunar distances above, except that one altitude observation only is needed and, in the reduction, only Step A was needed. This considerably reduced the labour involved; furthermore, the method is potentially far more precise.

II TIME-FINDING BY ASTRONOMY

The method

For astronomical purposes, the basic requirement in time-finding is the determination of the precise moment of *mean noon*, that is, the moment the Mean *Sun* is at its highest, or *culminates*, which occurs when it crosses, or *transits*, the observer's meridian. The interval between two successive transits of the Mean Sun across the same meridian is a *Mean Solar day*, which can then be divided into *hours*, *minutes*, and *seconds* by some form of timekeeper such as a sundial or clock. The Mean Sun is a fictitious body, created by astronomers when it was realized in Hellenistic times that the real Sun was not the best of timekeepers, sometimes running a little fast, sometimes a little slow. Of course, until the coming of accurate clocks in the last few hundred years, the concept of *mean solar time* was of interest only to astronomers: for the purposes of ordinary life, it was the real Sun – its rising, its culmination, its setting – which governed the time of day, and it was the sundial which told the time – *apparent solar time* – when needed.

Here, however, we *are* concerned with astronomers and therefore with mean rather than apparent time. The Mean Sun being a fictitious body which moves around the celestial equator at a constant speed (whereas the real Sun moves around the ecliptic at varying speeds in the course of the year), the moment of mean noon cannot be determined by direct observation but only indirectly through observations of some real body. For obvious reasons, it was observations of the real Sun which were used for this purpose in earlier times, the instant of mean noon being then obtained by applying the *equation of time*, which is the difference between mean and apparent solar time at any moment. Depending as it does upon the Sun's declination north or south of the celestial equator, and upon the Sun-Earth distance on the day concerned, the size of the equation varies through the year. On 4 November each year, for example, the real Sun crosses the meridian some 16 minutes before mean noon, while on 2 September the equation is zero and mean and apparent times are the same.

But using the real Sun for precise time determination presents a number of disadvantages: the Sun is not easy to observe accurately and it may be cloudy at the critical moment. Astronomers, therefore, seldom use the Sun for time determination, preferring instead to observe certain bright stars whose positions are very accurately known. Direct observations of any one of these will yield *sidereal time*, which can be converted to mean solar time whenever needed. The position of a star which appears to the observer as a point of light can be determined very precisely. As there are many stars, cloudy weather is less of a hazard and the overall precision of time determination is increased because a dozen or more observations can be taken each night (and, for the brighter stars, in daylight as well), whereas it is generally not possible to take more than one solar observation for time determination in any one day.

The instruments

At the time Greenwich Observatory was founded the usual method of time-finding was by the Double Altitude, or Equal Altitude, Method, not to be confused with the later Gaussian method which is sometimes called by the latter name. In the Double Altitude Method the astronomer, using a movable quadrant mounted on a vertical axis, took an altitude observation of the Sun when the altitude was increasing, an hour or so before noon, noting the time by the clock being checked. Then, at an equivalent interval after noon – the Sun having crossed the meridian and with its altitude now decreasing – he noted the precise time the Sun reached the same altitude as in the forenoon observation. With a small correction to allow for the change in the Sun's declination between observations, the time halfway between the two equal-altitude observations was apparent noon, from which mean noon could be found by applying the equation of time for the day. This was the principal method used at Greenwich from 1676 to 1725, the various quadrants being set up in the Great Room.

The invention of the transit instrument, which in its various forms was used for time determination at the Royal Observatory from 1721 to 1957, is ascribed to Ole Römer, the Danish astronomer who first used such an instrument in Copenhagen in 1689. The first to be used in England was set up by Edmond Halley in 1721 at Greenwich, where it can still be seen. The simple transit instrument consists of a telescope fixed at right angles to a horizontal axis which is free to rotate upon pivots on two fixed piers. The telescope can be moved up and down but not from side to side and, because the axis is adjusted so as to be horizontal and precisely east and west, the line of sight of the centre of the field of view (the technical term is line of collimation, or optical axis) will always be in the plane of the observer's meridian. In principle, therefore, the instant any heavenly body crosses the meridian can be found from a single observation. In practice, the transit telescope has several horizontal and vertical 'wires' – actually they are usually spider's threads set at the telescope focus – which can be seen in the field of view. It is essential that the optical axis of the transit telescope should always be directed towards the meridian. To ensure this, the transit observer has to take account of three possible errors, preferably before each set of observations – either to adjust the apparatus to remove them, or to know their magnitude so that they can be allowed for arithmetically:

(a) *azimuth error*, when the pivots on which the axis rotates are not exactly east and west: checked by observing an azimuth mark set precisely on the meridian a mile or so distant (or by observing circumpolar stars) and corrected by moving one of the pivots backwards or forwards;

(b) *level error*, when the pivots are not horizontal: checked by a bubble level or nadir observations (looking vertically downwards into a bowl of mercury) and corrected by raising or lowering one of the pivots;

(c) *collimation error*, when the optical axis of the telescope is not at right angles to the rotation axis of the pivots: eliminated if the telescope can be reversed on its bearings during an observation.

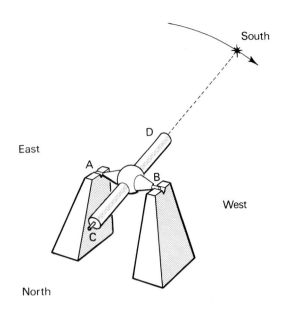

A transit instrument, used in conjunction with a clock to determine the exact moment when a heavenly body crosses the meridian. The axis *AB* is horizontal and points east and west, so that as the telescope *CD* swings on the axis it moves in the meridian.

Simple transit instruments were used for time-finding at Greenwich from 1721 to 1850[1] but it must be remembered that the transit instrument (and its development, the transit circle) is used extensively for the converse problem: given the time, find the right ascension.

Airy's transit circle which can measure declination as well as right ascensions and whose optical axis defines the world's Prime Meridian was used for time-finding from 1851 to 1927. Time was recorded automatically by chronograph from 1854, and many other refinements were introduced over the years. However, the increasing precision in timekeeping brought about by the free-pendulum clocks gave rise to the need for a corresponding increase in precision in time determination. The telescope and axis of Airy's transit circle weigh nearly two thousand pounds and cannot be quickly reversed on the pivots. From 1927, therefore, time determination (but not position measurement) at Greenwich (and during and after the Second World War, at Abinger, Edinburgh, and Herstmonceux as well) was taken over by a succession of small reversible transit instruments (with telescopes about three feet long) on special stands which allowed the transit telescope to be reversed during an observation, thus eliminating collimation error.

But, as we have seen, timekeeping precision increased fast in the 1940s and 1950s, first with the quartz crystal and then with the atomic clock, demanding yet more precision in time determination. In 1957, the observatory having moved to Herstmonceux, time determination was taken over a Photographic Zenith Tube (PZT) designed by D. S. Perfect.[2] The PZT is a development of the Reflex Zenith Tube devised by Airy about 1850. Airy's design was subsequently greatly improved

Left. The Greenwich photographic zenith tube (PZT), 1974. Right. The Greenwich PZT Console. There is no 'observer', as such, the whole operation of the zenith tube being remotely controlled from this console.

and adapted for use in time determination at the US Naval Observatory. As the name implies, the telescope of the PZT is fixed so as to point vertically towards the zenith (through which the meridian passes). Its use is limited to measuring the transits of stars passing within $15'$ of the zenith but this drawback is offset by the superior accuracy of the measurements. Rays from a star are reflected upwards from the mercury surface which defines the vertical, and the images from four successive exposures are automatically recorded on a moving photographic plate set centrally under the lens. The clock times of each exposure are accurately (and automatically) recorded by chronograph and the times at which the stars transit the meridian can be obtained from the positions of the images on the plate and the clock times at which they are observed. To obtain GMT (strictly, UTo), the local time obtained has to be corrected for the amount Herstmonceux is to the east of Greenwich – $1^m\,21^s.0785$.

Another instrument that has been used regularly at the RGO, and used extensively for time determination in many observatories, is the impersonal prismatic astrolabe invented by André Danjon and used for the first time in 1951 at Besançon.[3] It can determine very precisely when a star reaches a fixed zenith distance of $30°$ and it has a rotary mounting so that it can be pointed in any direction. As with the PZT, the vertical is defined by reflection from a bowl of mercury. The transit instrument and PZT make use of the apparent motion of the stars across the meridian, whereas the prismatic astrolabe makes use of the motion of several stars across the almucantar (line of equal zenith distance) of 30°, the time being found by the equal-altitude method, invented by Gauss in 1808.[4]

Over three hundred years ago, Flamsteed's double-altitude time-determination observations were probably accurate to approximately 5 seconds of time. At the beginning of the present century, the accuracy expected in time determination with a conventional transit instrument was about 1/10 second (100 milliseconds). In the 1970s the PZT measures to 20 milliseconds on observations of a single star, or to 4 milliseconds on observations of, say, 30 stars - an exercise which can mean a full night's work.

All of the astronomical instruments for monitoring the rotation of the Earth described so far have used optical observations, but these are now all obsolescent. Developments in both radio astronomy and satellite tracking since the 1970s have resulted in methods many times more precise, so that the position of the pole and the rotation of the Earth can be measured to an accuracy of a centimetre.

The radio astronomy method is known as Very Long Baseline Interferometry (VLBI), involving simultaneous observations of radio waves from Quasars – the most luminous and distant objects visible in the universe – by radio telescopes separated by distances up to 1,000 km. The relative time of arrival of the radio waves at the various VLBI radio telescopes gives a measurement of the rotation of the Earth and the position of the pole at that distance to the nearest centimetre.

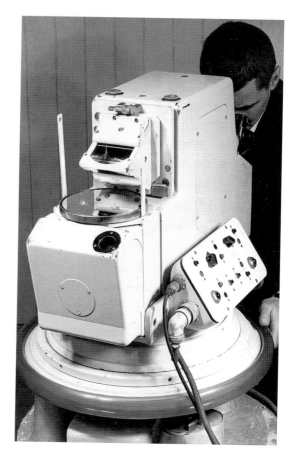

A Danjon impersonal prismatic astrolabe at Paris Observatory.

The satellite tracking method involves making range measurements – to better than 1 cm – either of special artificial satellites equipped with retro-reflectors (Satellite Laser Ranging, or SLR), or of retro-reflectors placed on the Moon by the US and Russian space missions (Lunar Laser Ranging, or LLR). This is done by making precise measurements of the time of flight of very short laser pulses from a laser ranging station on Earth to the satellite and back.

From these range measurements, very precise details of the rotation of the Earth and the motion of the pole can be determined, again using the laser ranging data. The accuracy is similar to that achieved from VLBI, with SLR having the capability of giving a very rapid determination of Earth rotation, but lacking the high stability of the VLBI frame over longer time periods.

III Greenwich meridian in the space age

Time itself has a unique scale and an origin. For most of us, the scale is Atomic Time (TAI, see p. 174). Universal Time (UT) is a defined time, synonymous with Greenwich Mean Time, which it replaces; but Greenwich is certainly not forgotten as scientists explore time and space.

Sidereal, radio and atomic time

Time can be kept with reference to the rotation of the Earth and this type of time is called Sidereal time. Universal Time 1 (UT1) is related to the Earth rotation time, which is still called Greenwich Mean Sidereal Time (GMST). GMST is now referred to the International Meridian (IRM), defined within the International Celestial Reference System (ICRF), rather than the Greenwich Meridian.

Civilian time (Broadcast), as recommended by the International Radio Consultative Committee (CCR), is Universal Time Co-ordinated (UTC). This is the time that we hear on radio transmissions.

UTC is related to Atomic Time (TAI) by a correction defined by the International Earth Rotation Service (IERS). The correction is given as a whole second. So, the difference between UT1 and TAI includes fractions of a second, but the difference between TAI and UTC does not (30 seconds, as of 1 Jan. 1996).

The Global Positioning System, GPS, has its own GPS time, kept by atomic clocks so it uses the TAI time-scale, which was identical to UTC on 5 January 1980. GPS observations enable positions and time to be determined anywhere in or around the Earth.

Longitude and Airy's Transit position

The link between longitude and time is defined by the IERS Terrestrial Reference Frame (ITRF). This is based on observations to satellites and celestial compact radio sources (quasars) from co-ordinated stations around the globe.

In Europe ITRF is realized as ETRF, the European Terrestrial Reference Frame. In 1989 they were identical, but are slowly moving apart due to tectonic plate movements and inconsistencies in the movement of the Earth in its orbit and rotation around its axis. Because of small, but observable movement of a centimetre or so, reference frames are suffixed by a date. In 1989, ITRF89 and ETRF were identical. Since then ETRF has moved with the stable part of Europe. This is a small movement, a few centimetres relative to ITRF.

Comparing the longitude of the Airy Transit in the system available in 1936 to the longitude determined with space techniques gives a difference.

Using Ordnance Survey Level 1 Transformation

Airy Transit, GB36 =	N 51 28 38.265	E 00 00 00.418
Airy Transit, ETRF89 =	N 51 28 40.1247	W 00 00 05.310
	X 3980637.8044	Y -102.4779 Z 4966897.8318

The difference in longitude between the two systems at the Airy Transit is 102.478 metres. Therefore, the International Reference Meridian is 102.5 metres west of the Airy Transit at Greenwich, which will be the real reference meridian to be used for the Millennium.

Carl Calvert, Ordnance Survey

Bibliography

Andrewes, W. G. H. (ed.), *The Quest for Longitude* (Cambridge, Mass., 1996).

Baily, F., *An Account of the Revd. John Flamsteed.* (London, 1835).

Bartky, Ian R., 'The Bygone Era of Time Balls', *Sky & Telescope*, January 1987, 32-5

Bartky, Ian R., 'The Adoption of Standard Time', *Technology & Culture*, Jan. 1989, 25-56

Bigourdan, G., 'Le Jour et ses divisions. Les fuseaux horaires et l'Association Internationale de l'Heure', *Annuaire du Bureau des Longitudes* (Paris, 1914).

Bigourdan, G., 'Les services horaires de l'observatoire de Paris. . .', *Bulletin Astronomique*, II [1921-2].

Blair, B. E. (ed.), *Time and Frequency: Theory and Fundamentals* (US National Bureau of Standards, May 1974).

Brown, L. A., *The Story of Maps* (New York, 1951).

Chapin, Seymour, 'A survey of the efforts to determine longitude at sea, 1660-1760', *Navigation*, 3, 7 (March 1953).

Corliss, C. J., *The Day of Two Noons* (Washington, DC, 1941).

Cotter, C. H., *A History of Nautical Astronomy* (London and Sydney, 1968).

De Carle, D., *British Time* (London, 1947).

Ditisheim, P., *et al.*, *Pierre Le Roy et la Chronomètre* (Paris, 1940).

Dowd, Charles N. (ed.), *Charles F. Dowd, A.M., Ph.D. and Standard Time* (New York: Knickerbocker Press, 1930).

Ellis, William, 'Lecture on the Greenwich System of Time Signals', *The Horological Journal*, 1 May 1865.

Essen, L., *The Measurement of Frequency and Time Interval* (HMSO: London, 1973).

Flamsteed, J., *The Correspondence of John Flamsteed, the First Astronomer Royal*, ed. E. G. Forbes, L. Murdin & F. Willmoth, Vol. I, 1666-1682 (Bristol & Philadelphia, 1995).

Forbes, E. G., *The Birth of Navigational Science* (Greenwich, 1974).

Forbes, E.G., *Greenwich Observatory, vol. i: Origins and Early History* (London, 1975).

Forbes, E. G., 'The Origins of the Royal Observatory at Greenwich', *Vistas in Astronomy*, 20 (1976).

Gould, R. T., *The Marine Chronometer: its history and development* (London, 1923).

Guyot, E., *Histoire de la détermination des longitudes* (La Chaux-de-Fonds, 1955).

Guyot, E., *Histoire de la détermination de l'heure* (La Chaux-de-Fonds, 1968).

Haswell, J. E., *Horology* (London, 1976).

Hope-Jones, F., *Electrical Timekeeping* (London, 1976).

Howse, Derek, *Greenwich Observatory, vol. iii: Its Buildings and Instruments* (London, 1975).

Howse, Derek, and Hutchinson, B., *The Clocks and Watches of Captain James Cook, 1769-1969* (Antiquarian Horology reprint, 1969).

Howse, Derek, *Nevil Maskelyne: The Seaman's Astronomer* (Cambridge, 1989).

Howse, Derek, and Hutchinson B., *The Tompion clocks at Greenwich and the dead-beat escapement* (Antiquarian Horology reprint, 1970-1).

Howse, Derek, 'Le Bureau britannique des Longitudes', *L'Astronomie*, Oct. 1978, 413-25.

Jespersen, J., and Fitz-Randolph, J., *From Sundials to Atomic Clocks: understanding time and frequency*, National Bureau of Standards Monograph, 155 (Washington, DC, 1977).

Kieve, J. L., *Electric Telegraph: A Social and Economic History* (Newton Abbot, 1973).

Landes, David S., *Revolution in Time: Clocks and the Making of the Modern World* (Cambridge, Mass., 1983).

Leigh-Browne, F. S., 'The International Date Line', *The Geographical Magazine*, April 1942.

McCrea, W. H., *The Royal Greenwich Observatory* (London, 1975).

Maindron, Ernest, *Les Fondations de Prix à l'Académie des Sciences - Les Lauréats de l'Académie, 1714-1880* (Paris, 1881).

Malin, Stuart, & Stott, Carole, *The Greenwich Meridian* (Southampton, 1984).

Malin, Stuart R., Roy, A. E., and Beer P., (eds.), 'The Longitude Zero Symposium 1984', *Vistas in Astronomy* 28, 1/2 (1985).

Marguet, F., *Histoire de la longitude à la mer au XVIII° siècle, en France* (Paris, 1917).

Maunder, E. W., *The Royal Observatory, Greenwich* (London, 1900).

May, W. E., 'How the chronometer went to sea', *Antiquarian Horology* (March 1976).

Mayall, R. Newton, 'The Inventor of Standard Time', *Popular Astronomy*, L, no.4 (April 1942).

Morando, B., 'Le Bureau des Longitudes', *L'Astronomie*, 90 (June 1976).

Neugebauer, O., *A History of Ancient Mathematical Astronomy* (Berlin, Heidelberg and New York, 1975).

Observatoire de Paris, *75 ans au service de l'heure universelle* (Paris, 1992)

Pedersen, O., & Pihl, M., *Early Physics and Astronomy: a historical introduction* (1974).

Perrin, W. G., 'The Prime Meridian', *Mariner's Mirror*, XIII, no.2 (April 1927).

Quill, Humphrey, *John Harrison, the Man who found Longitude* (London, 1966).

Randles, W. G. L. , 'Portuguese and Spanish attempts to measure longitude in the sixteenth century', *Mariners Mirror*, Vol. 81, No. 4 (Nov. 1995), 402-8.

Sadler, D. H., *Man is not Lost: a record of two hundred years of astronomical navigation with the Nautical Almanac, 1767-1967* (London, 1968).

Sadler, D. H., 'Mean Solar Time on the Meridian of Greenwich', *Quarterly Journal of the Royal Astronomical Society*, 19, (1978).

Smith, H. M., 'International time and frequency coordination', *Proceedings of the IEEE*, 60, no.5 (May 1972).

Smith, H. M. , 'The Bureau International de l'Heure', *Proceedings of the 8th Annual PTTI Applications and Planning Meeting*, Nov.-Dec. 1976.

Smith, H. M., 'Greenwich time and the prime meridian', *Vistas in Astronomy*, 20 (1976).

Sobel, Dava, *Longitude. The true story of a lone genius who solved the greatest scientific problem of his time* (London, 1995).

US Government, *International Conference held at Washington for the purpose of fixing a Prime Meridian and a Universal Day, October 1884 - Protocols of the Proceedings* (Washington, DC, 1884).

Waters, D. W., *The Art of Navigation in England in Elizabethan and early Stuart times* (1958, reprinted Greenwich, 1978).

Weber, Gustavus A., *The Naval Observatory, its History, Activities and Organization* (Baltimore, 1926).

Wheeler, G. M., *Report upon the Third International Geographical Congress and Exhibition at Venice, Italy, 1881. . .* (Washington, DC, 1885).

Notes

Full titles and sources are given only for those works not listed in the Bibliography (pp. 190-1).

Introduction (pp. 12–15)

1 US Government, *International Conference held at Washington October 1884 - Protocol of the Proceedings*, 199-201.
2 ibid. 199.

Chapter 1 (pp. 17-31)

1 Neugebauer, O., *A History of Ancient Mathematical Astronomy*, 1934.
2 Stevenson, E. L. (trans. & ed.), *Geography of Claudius Ptolemy* (New York, 1932), 28.
3 Neugebauer, op. cit. 938.
4 Taylor, E.G. R., *The Haven-finding Art*, 55.
5 Preserved in the Bibliothèque Nationale, Paris (called the *Carte Pisane*, as it is said to have been acquired from a Pisan family).
6 As for example *Opus Almanach magistri Johānis de monte regio ad annos xviii explicit felicit. Erhardi Radolt Augusten Vindelicorum ... Septembris M. cccc. lxxxviij* [1488]. This has almanacs for the years 1489 to 1506 inclusive and is likely to have been the edition carried by Columbus.
7 *Almanach noua plurimis annis venturis inseruientia: per Ioannem Stoefflerinum Instingersen & Iacoben Pflaumen Vlamensem accuratissime supputatia: & toti fere Europe dextro sydere impartita* [1499].
8 Morison, S. E., *Admiral of the Ocean Sea* (Boston, 1942), II. 158-9, 162.
9 ibid. II. 400-3, 406.
10 Randles 403.
11 Werner, J., *In hoc opere haec ōtinentur Noua translatio primi libri geographicae Cl'Ptolomaei ...* (Nuremberg, 1514), Cap. IV, Annot. 8.
12 Pedersen, O., & Pihl, M., *Early Physics and Astronomy*, 258.
13 Werner, op. cit. sig. dvʳ.
14 Apian, Peter, *Cosmographicus Liber Petri Apiani Mathematici Studiose Collectus* (Ingolstadt, 1524), ff. 30-1.
15 Gemma Frisius, *Cosmographicus Liber Petri Apiani Mathematici, iam de nouo integritati restitutus per Gemmam Phrysium* (Antwerp, 1533), ff. xvᵛ~xviʳ.
16 Cuningham, William, *The Cosmographical Glasse, conteinyng the pleasant Principles of Cosmographie, Geographie, Hydrographie, or Nauigation* (London, 1559), f. 107.
17 Gemma Frisius, *Gemma Phrysius de Principiis Astronomiae & Cosmographiae ... vsv Globi et eodem editi* (Antwerp, 1530), sigs. D2ᵛ–D3ˣ. (Translation by Philip Kay.)
18 Gemma Frisius, *De Principiis Astronomiae . . .* (Antwerp, 1553), 65. (Translation by Philip Kay.)
19 Eden, Richard, *The Decades of the New Worlde ...* (London, 1555), f. 361.
20 Cuningham, op. cit., f. 110.
21 For example, Cervantes, Miguel de, 'The famous adventures of the enchanted bark', *Don Quixote*, 2nd part, ch. XXIX.
22 Cervantes, Miguel de, *El Coloquio de los Perros*, in *Obras Completas* (Madrid, 1944), 244.
23 Gould, R. T., *The Marine Chronometer*, 12.
24 Marguet, F., *Histoire de la Longitude*, 45.
25 Brown, L. A., *The Story of Maps*, 209.
26 Dreyer, J. L. E., 'Time, Measurement of', *Encyclopaedia Britannica*, 11th edn., XXVI (1911), 984.
27 Kepler, Johannes, *Tabulae Rudolphinae ...* (1627).

28 Marguet, op. cit. 7.
29 Guyot, E., *Histoire de la détermination des longitudes*, 11.
30 Tarpalling = tarpaulin = sailor (tar).
31 Sherwood Taylor, F., 'An early satirical poem on the Royal Society', *Notes and Records of the Royal Society*, Oct.1947, 37-46.
32 Histoire de l'Académie Royale des Sciences, I (1668), 67-9. (There is a fuller account in Huygens, Christiaan, Oeuvres . . ., XXII, 218-26.) (Translations by Dr Barbara Haines.).
33 ibid. 67.

Chapter 2 (pp. 33-51)

1 *Dictionary of National Biography*, XIII, 820-1.
2 Lawson Dick, O., *Aubrey's Brief Lives* (Penguin English Library, 1972), 370-1.
3 Plumley, N., 'The Royal Mathematical School within Christ's Hospital', *Vistas in Astronomy* (1976), vol.20, pp.51-9.
4 Baily, F., *An Account of the Revd. John Flamsteed ...*, 29-31. Henceforth cited as Baily.
5 PRO State Papers Domestic, Entry Book 27, f. 59.
6 Baily, 37-8.
7 Forbes, E. G., 'The Origins of the Royal Observatory at Greenwich', p.48, note 17, discusses St. Pierre's identity.
8 British Library Add. MS. Birch 4393, f. 89. Holograph copy with signature of King and Williamson (PRO/SP44/334/27-8).
9 Taylor, E. G. R., 'Old Henry Bond and the Longitude', *Mariner's Mirror*, 25 (1939), 162–9.
10 Flamsteed to Sherburne, 12 July 1682 (Baily, 125-6).
11 Memorandum by Pell, 1675 (BL Add. MS. Birch 4393, f. 93ʳ).
12 Flamsteed, J., *Historia Coelestis Britannica*, III (London, 1725), Prolegomena, 102 (translated by Mrs E. M. Barker).
13 Flamsteed to Sherburne, 12 July 1682 (Baily, 126).
14 Baily, 112. Copies in PRO/SPDom.29/368, f. 299 and SP Dom.44/44, p.10.
15 Baily, 37, note.
16 Baily, 39.
17 Memorandum by Pell, 1675 (op. cit. f. 93v).
18 Forbes, E. G., 'The Origins of the Royal Observatory', 39-48.
19 Williamson to Pell, 23 April 1675 (BL Add. MS Birch 4393, f. 97).
20 Flamsteed to Sherburne, 12 July 1682 (Baily, 126).
21 PRO/SP Dom.44, 15 (Baily, 112).
22 Howarth, W., *Greenwich: past and present* (London & Greenwich, c. 1886), 84.
23 *A Mapp or Description of the River of Thames ... made by Jonas Moore Gent: ... 1662*. Pen, watercolour, gold paint on vellum. Museum of London on loan from PRO. Reproduced in Howse, *Greenwich Observatory*, iii, fig. 1.
24 SP Dom. 15 (Baily, 112).
25 Howse, Derek, *Greenwich Observatory, vol. iii: Its Buildings and Instruments*, 5.
26 PRO/WO.47/19b.
27 Howse, Derek, and Hutchinson, B., *The Tompion clocks at Greenwich and the dead-beat escapement*, 24.
28 Wren to Fell, 3 Dec. 1681 *(Wren Society*, V (Oxford, 1928), 21-2).
29 R. Society MSS. 243 (F1).
30 ibid.
31 Cambridge University Library MS RGO 1/1 f.22ʳ. Henceforth RGO references will omit 'CUL MS'.
32 Horrox, J., *Opera Postumia... in calce adjiciuntur Johannis Flamsteedii, Derbiensis, de Temporis Aequatione Diatriba...* (London, 1673).
33 R. Society MSS. 243 (Fl).

34 RGO 1/36, f.54ᵛ.
35 Howse, Derek, op. cit. note 27.
36 Baily, 99.
37 Flamsteed to Sharp, 29 March 1716 (Baily, 321).
38 ibid.
39 Chapman, Alan (ed.), *The Preface to John Flamsteed's* Historia Coelestis Britannica *...(1725)* (Greenwich, 1982).

Chapter 3 (pp. 45–80)
1 May, W. E., 'The Last Voyage of Sir Clowdisley Shovel', *J. Inst. Navig.* 13 (1960), 324-32.
2 Lanoue, Capt. H., memorial of 1736, cited by Moriarty, H. A., in 'Navigation', *Encyclopaedia Britannica*, 9th edn. (1884), XVII, 258.
3 May, op. cit.
4 *The Guardian*, no.107 (14 July 1713), 254-5.
5 Osborn, James M., '"That on Whiston" by John Gay', *Bibliographical Society of America, Papers*, lvi (1962), 73.
6 *The Guardian*, loc. cit. 255-6.
7 Whiston, W., and Ditton, H., *A New Method for Discovering the Longitude both at Sea and on Land* (London, 1714).
8 Osborn, op. cit. 74.
9 Swift, Arbuthnot, Pope, and Gay, *Miscellanies, the Fourth Volume. Consisting of Verses by Dr. Swift, Dr. Arbuthnot, Mr. Pope, and Mr. Gay* (London, 1747), 145-6.
10 Rawson, C. J., 'Parnell on Whiston', *Bibliographical Society of America, Papers*, lvii (1963), 91-2, citing BL Add. 38157.
11 *House of Commons Journal*, 17 (25 May 1714).
12 ibid. (11 June 1714).
13 Act 12 Anne *cap.* 15 (13 Anne *cap.* 14 by modern notation). Quoted in full in Quill, Humphrey, *John Harrison, the Man who found Longitude*, 225-7.
14 Landes, 112.
15 Gould, *The Marine Chronometer*, 32-5.
16 Hobbs, William, Broadsheet dated 15 Sept. 1714 in RGO 1/69, f. 257ʳ.
17 Bull, Digby, Letter to Commissioners dated 29 Sept. 1714, RGO 1/36, f. 116ʳ.
18 Brewster, Sir David, *Memoirs of the Life, Writings, and Discussions of Sir Isaac Newton* (Edinburgh, 1855; Johnson repr., 1965), II. 263.
19 Brewster, Sir David, 'On Sir Christopher Wren's Cipher, containing Three Methods of finding the longitude', *Report of the Twenty-ninth meeting of the British Association . . . held at Aberdeen September 1859* (1860), first pagination, 34. The original cryptogram at the Royal Society has been mislaid but a copy is said to be among the Portsmouth papers. It seems likely that the Brewster version may have contained some mistakes.
20 Bennett, J. A., 'Studies in the Life and Work of Christopher Wren'. PhD thesis, Cambridge University, 1974, 263-5.
21 Swift, Jonathan, *Gulliver's Travels* (1726), Dent, The Children's Illustrated Classics (1952), 202.
22 Paulson, Ronald, *Hogarth's Graphic Works* (revised edn., London and New Haven, 1970), I.169-70.
23 Goldsmith, Oliver, *She Stoops to Conquer* (1773), Act I.
24 Maindron, E., *Les Fondations de Prix à l'Académie des Sciences*, 15.
25 ibid. 23.
26 ibid. 17.
27 Marguet, F., *Histoire de la longitude*, 85-7.
28 Chapin, Seymour, 'A survey of the efforts to determine longitude at sea, 247.
29 Brown, L. A., *The Story of Maps*, 186-90.
30 *Phil. Trans.* 37, no.420 (1731), 145-57.
31 *Dictionary of American Biography*, IV, 345-6.
32 Bedini, Silvio, *Thinkers and Tinkers: Early American Men of Science* (New York, 1975), 118.
33 Shepherd, A., *Tables for Correcting the Apparent Distance of the Moon and a Star from the Effects of Refraction and Parallax*

(London, 1772), Preface.
34 Halley, E., 'A Proposal of a Method for finding the Longitude at Sea within a Degree, or twenty Leagues', *Phil. Trans.* 37, no.421 (1731), 195.
35 Chapin, Seymour, 'A survey of the efforts to determine longitude at sea', 247-8.
36 RGO 14/5 pp. 20–21.
37 Maskelyne, N., *The British Mariners' Guide* (London, 1763).
38 Forbes, E. G., *Greenwich Observatory, vol. i: Origins and Early History*, 120-1.
39 RGO 14/5, p. 51 (7 Aug. 1763).
40 RGO 14/5, pp. 79–80 (9 Feb. 1765).
41 ibid.
42 Nevil Maskelyne to Edmund Maskelyne, 15 May 1766 (NMM MSS. PST/76, ff. 104-6).
43 Howse, 'The Lunar-distance Method ...', in Andrewes (ed.) *Quest* (1996), 150 ff.
44 Royal Warrant of 4 March 1674/5.
45 *The Nautical Almanac and Astronomical Ephemeris for the Year 1767* (1766).
46 For example: Gould, R. T., *The Marine Chronometer*; Quill, Humphrey, *John Harrison, the Man who found Longtitude*; Sobel, *Longitude*. The best contemporary account for the earlier history was: Anon. [probably James Short and Taylor White; see n. 58, below], *Account of the Proceedings in Order to the Discovery of the Longitude at Sea* (London, 1763).
47 RGO 14/5, p. 24.
48 Act 3 Geo. III, *cap.* 14.
49 RGO 14/5, p. 60.
50 Act 5 Geo. III, *cap.* 20.
51 Howse, D., 'Captain Cook's marine timekeepers', *Antiquarian Horology* (1969), 190-205, reprinted as *The Clocks and Watches of Captain James Cook, 1769-1969*.
52 Beaglehole, J. C., *The Journals of Captain James Cook*, II (London, 1961), 692.
53 Wales, W., & Bayly, W., *The Original Observations made . . . in the years 1772, 1773, 1774 and 1775 ...* (London, 1777), 280.
54 Betts, J., 'Arnold and Earnshaw ...', in Andrewes, op. cit. 312 ff.
55 May, W. E., 'How the chronometer went to sea', 638-63.
56 Cardinal, C., 'Ferdinand Berthoud and Pierre Le Roy', in Andrewes, op. cit., 282 ff.
57 Nivernois to Praslin, 21 March 1763. The subsequent story is told in the following manuscripts of the Académie des Sciences, Paris: Praslin to Choiseul, 28 March 1763; Choiseul to Académie, 31 March 1763; Académie to Choiseul, 4 April 1763; extract from Académie register, 16 April 1763.
58 Chapin, Seymour, 'Lalande and the longitude: a little-known London voyage of 1763'; *Notes and Records of the Royal Society*, 32 (1978).
59 Camus to Morton, 2 June 1763; Morton to Camus, 3 June 1763; contemporary copies among the papers of James Stuart Mackenzie, Mount Stuart, Rothesay, Bute, quoted by permission of the Marquess of Bute.
60 Le Roy, Pierre, *Exposé succinct des Travaux de MM. Harrison et Le Roy dans la Recherche des Longitudes en Mer & des épreuves faites de leurs Ouvrages* (Paris, 1768), 34-5.
61 Ditisheim, P., *et al.*, *Pierre Le Roy et la Chronomètre*, 100-1, quoting from Berthoud, *Traité des Montres à Longitude ...* (Paris, 1792), and Berthoud to Minister of Marine, 26 Dec. 1765 (Bibl. Nat.; Nouv. acq. français 9849).
62 Maindron, Ernest, *Les Fondations de Prix. . .*, 21.
63 Marguet, F., *Histoire de la longitude*, and Guyot, F., *Histoire de la détermination des longitudes*, both give excellent reviews of the early history of the French chronometer makers.
64 *Lois, Décrets, Ordonnances et Décisions concernant le Bureau des Longitudes* (Paris, 1909), 1-15.

65 Morando, B., 'Le Bureau des Longitudes', 279-94.
66 Bartky, Ian R., & Dick, Steven J., 'The first Timeballs', *JHA* xii (1981) 155-64.
67 *Nautical Magazine* (London, 28 Oct. 1833), 680.
68 See: 'Neptune', 'The Time-ball at Greenwich', *Nautical Magazine*, IV (1835), 584-6; Laurie, P.S., 'The Greenwich Time-ball', *The Observatory*, 78, no.904 (June 1958), 113-15; Howse, Derek, *Greenwich Observatory, vol. iii: Its Buildings and Instruments*, 134-6.
69 Bartky, Ian R., 'The Bygone Era of Time Balls', *Sky & Telescope* Jan. 1987, 32-5; Bartky & Dick, 'The First North American Time ball', *JHA* xiii (1982) 50-5.

Chapter 4 (pp. 87–115)
1   An excellent historical description of how the day was divided by various peoples from the earliest times is contained in Bigourdan, G., 'Le Jour et ses divisions . . .', B1-B40. See also Neugebauer, O., *A History of Ancient Mathematical Astronomy*.
2   Bigourdan, op. cit. B8-9.
3   Joyce, H., *The History of the Post Office* (1893), 283.
4   RGO 74. P. J. Melotte papers.
5   Airy, G. B., *Report of the Astronomer Royal to the Board of Visitors* (1856/7), 15. Henceforward cited as *Report*.
6   *Illustrated London News*, 14 May 1842, 16.
7   PRO/RAIL 1005/235, f. 58, 3 Nov. 1840.
8   B. L. Vulliamy, 'On the Construction and Regulation of Clocks for Railway Stations', *Proc. ICE*, p. 15 of offprint.
9   PRO/RAIL 1008/95.
10 Booth, Henry, *Uniformity of Time, considered especially in reference to Railway Transit and the Operations of the Electric Telegraph* (London and Liverpool, 1847), 4.
11 ibid. 16.
12 Smith, H. M., 'Greenwich time and the prime meridian', 221.
13 PRO/RAIL 1007/393 (for L & NWR) and private communication from Prof. J. Simmons (for Caledonian R.).
14 *Illustrated London News*, 13, 23 Dec. 1848, 387.
15 Bagwell, P., *The Transport Revolution from 1770* (London, 1970).
16 *Manchester Guardian*, 1 December 1847, 1.
17 Airy, *Report* (1849), i6.
18 RGO, MSS. 6/610.
19 Kieve, J. L., *Electric Telegraph: A Social and Economic History*, 104.
20 RGO, 6/610 and 611 for correspondence with SER, ETC, and Admiralty; 6/724, section 37, and 6/725, section 49, for correspondence with Shepherd.
21 Airy's journal (RGO 6/25).
22 Ellis, William, 'Lecture on the Greenwich System of Time Signals', 98-9.
23 Varley, C. F., *Description of the Chronopher ...* Pamphlet published by the Electric and International Telegraph Company about 1864 (GPO Post Office Records - Post 81/46).
24 *The Times*, 19 June 1852.
25 Edwin Clark to Airy, 28 Aug. 1852, RGO 6/611, f.169ʳ.
26 RGO 6/611, section 1.
27 Cooper, B. K., and Lee, C. E., 'Standard or Zone Time', *The Railway Magazine*, Sept. 1935, 159.
28 RGO 6/610, section 3.
29 Baldock to Admiralty, 13 April 1852, RGO 6/619, section 1.
30 Beresford, C. F. C., & Cambridge, J. H., 'The Deal Time Ball', *Antiquarian Horology*, Autumn 1990, 33-43.
31 Unless otherwise stated, sources from this table are from (a) Ellis, W., op. cit. (n. 22), April-July 1865, 85-91, 97-102, 109-14, 121-4; (b) Ellis, W., 'Time signalling: a retrospect', *The Horological Journal*, Oct. 1911, 21-3; (c) Airy, *Report* (1868), 22.
32 Airy, *Report* (1868), 22.
33 Airy to Latimer Clark of E&ITC, 6 May 1861 (RGO 6/615, section 1).

34 Bartky (1987), op. cit., note 69 of Chapter 4.
35 Airy, G. B., quoted in Ellis, 'Lecture on the Greenwich System of Time Signals', 323.
36 *Manchester Guardian*, 14 August 1872, 1.
37 Airy, *Report* (1874), 16-17.
38 GPO/Post 30/E 9195/1888, file 2.
39 R[ussell], W. J., *Abraham Follett Osler, 1808-1903* [1904].
40 *The Times*, 12 Jan. 1850.
41 Edinburgh Town Council, Minutes, 4 Jan. 1848.
42 *Illustrated London News*, 12, 12 Feb. 1848, 89.
43 Anon., 'Greenwich time', *Blackwood's Edinburgh Magazine*, 63 (March 1848), 354-61.
44 For example; F[rodsham], C[harles], *Greenwich time: the universal standard of time throughout Great Britain* (London, 1848), 11 pp.
45 Anon., 'Railway-time aggression', *Chambers Edinburgh Journal*, XV, no. 390, new series (21 June 1851), 392-5.
46 *Illustrated London News*, 3 Jan. 1852, 10.
47 *The Times*, 21 Nov. 1851, 3.
48 ibid. 17 Nov. 1851, 7.
49 Northcote to Airy, 10 Aug. 1852 (RGO 6/597, ff.116ʳ–117ᵛ).
50 Airy to Northcote, 11 Aug. 1852 (RGO 6/597, ff.118ʳ–121ʳ).
51 Airy, *Report* (1853), 8.
52 *The Western Luminary*, 31 Aug. 1852 (RGO 6/610).
53 Tucker to Airy, 29 Oct. 1852 (RGO 6/597, ff.133ʳ, 134ʳ).
54 Latimer, J., *The Annals of Bristol in the Nineteenth Century* (Bristol, 1887) and personal communication from G. Langley, County Reference Librarian, 13 Feb. 1978.
55 *Plymouth, Devonport and Stonehouse Herald*, 18 Sept. 1852, 8.
56 Shenton, Rita, private communication, 21 March 1978.
57 Jagger, Cedric, *Philip Paul Barraud* (London, 1968), 68-9.
58 Davies, Alun C., 'Greenwich and Standard Time', *History Today*, 28(3) (March 1978), 198.
59 *The English Reports*, CL VII, Exchequer Division, XIII (1916), 719.
60 ibid.
61 ibid.
62 *The Times*, 14 May 1880.
63 Act 43 & 44 Vict. *c*.9. Unfortunately, no copy of the Committee's report, nor a record of any debate in either House, seems to have survived (House of Lords Record Office).
64 Baker, E. C., 'Post Office clocks', *Post Office Telecommunications Journal*, Feb.1954, 54-5.

Chapter 5 (pp.117-43)
1   Forbes, E. G., *Greenwich Observatory, vol. i: Origins and Early History*, 150.
2   Airy, *Reports* (1855), 11 (1863), Appendix III, 19.
3   Airy, *Report* (1867), 20, and RGO 6/16, section 1.
4   Airy, *Report* (1867), 21.
5   Bigourdan, G., 'Le jour et ses divisions. . .', B43, 4*n*.
6   Corliss, C. J., *The Day of Two Noons*, 3.
7   Weber, Gustavus A., *The Naval Observatory*, 27-8.
8   Langley, S. P., 'On the Allegheny System of Electric Time Signals', *The American Journal of Science and Arts* (1872), 377-86.
9   Carson, Mrs. Ruth, private communication dated 11 Nov. 1968, quoting Dowd, Charles N. (ed.), *Charles E. Dowd*.
10 Dowd, op. cit. Plate VI.
11 (a) Smith, H. M., 'Greenwich time and the prime meridian', 222-3, also citing primary sources; (b) Dowd, op. cit.; (c) Mayall, R. N., 'The Inventor of Standard Time'; (d) Bartky (1989).
12 *New York Herald*, Sunday 18 Nov. 1883, 10.
13 ibid. 19 Nov. 1883, 6.
14 *Harper's Weekly* (New York), 29 Dec. 1883, 843.
15 *Popular Astronomy*, Jan. 1901.
16 *Detroit News*, 26 Sept.1938.
17 Rubio, José Pulido, *El Piloto Mayor de Ia Casa de la Contratación* (Sevilla, 1950), 438–41.

18 Perrin, W. G., 'The Prime Meridian', 118.
19 ibid. 119.
20 Anon., 'Remarques sur les Observations astronomiques faites aux Canaries en 1724 par le P. Feuillée, Minime', *Mém. de l'Acad. Royale des Sc. de Paris* (1742), 350-3.
21 Struve, Otto, 'The Resolutions of the Washington Meridian Conference', translation from German in Fleming, S., *Universal or Cosmic Time* (Toronto, 1885), 85 (in RGO Tracts, Geodesy, 02016(8)).
22 *Comptes-rendus des Congrès des Sciences Géographiques, Cosmographiques, et Commerciales tenu à Anvers du 14 au 22 Août 1871* (Anvers, 1882), 11. 254-5.
23 Borsari, Ferdinando, *Il Meridiano iniziale e l'ora universale* (Napoli, 1883), 61.
24 Smith, op. cit. 222.
25 Fleming, S., *Uniform non-local time (Terrestrial Time)* (Ottawa [1876]).
26 Mayall, R. Newton, 'The Inventor of Standard Time'.
27 Fleming S., *Time-reckoning and the selection of a prime meridian to be common to all nations* (Toronto, 1879).
28 Airy to Colonial Secretary, 18 June 1879, quoted in Fleming, S., 'Universal or Cosmic Time', *Proc. Canadian Inst.*, July 1885, 33.
29 Colonial Secretary to Governor General, Canada, 15 Oct. 1879, in Fleming, S., op. cit., note 28, 31.
30 Piazzi Smyth to Colonial Secretary, 5 Sept. 1879, in Fleming, S., *op. cit.*, note 28, 35-8.
31 'Memorandum of the Royal Society of Canada on the Unification of Time at Sea', *Trans. R. Soc. Canada* (1896-7), 11. 28.
32 De Beaumont, H. Bouthillier, *Choix d'un méridien initial unique* (Geneva, 1880).
33 Wheeler, G. M., *Report upon the Third International Geographical Congress ... 1881 ...* (Washington, DC, 1885), 28-9.
34 Smith, op. cit. 224-5.
35 *International Conference held at Washington for the purpose of fixing a Prime Meridian and a Universal Day, October 1884 - Protocols of the Proceedings.* Unless otherwise cited, all subsequent information on the Conference comes from this publication.
36 H[inks], A. R., 'Nautical time and civil date', *Geographical Journal*, LXXXVI, 2 (1935), 153-7.
37 Norie, J. W., *A New and Complete Epitome of Practical Navigation* (10th edn. 1831), 313.

Chapter 6 (pp. 145-61)
1 Pasquier, F., 'Unification of Time', quoted in *Journal of the British Astronomical Association*, Nov. 1891, 107.
2 Bigourdan, G., 'Le jour et ses divisions ...', B 60-8.
3 Décret no. 78-855 du 9 août 1978 relatif à l'heure légale française, *Journal Officiel de la République Française*, 19 Aug. 1978, 3080.
4 Bigourdan, op. cit. B 35.
5 'Memorandum of the Royal Soc. of Canada ...' (ch. 5, no.30), loc. cit. 15.
6 ibid. 48.
7 Bigourdan, op. cit. B 36.
8 Minutes of Conference on time-keeping at Sea, London, June–July 1917 (MOD, Hydrographic Dept. MSS.).
9 Sadler, D. H., 'Mean Solar Time on the Meridian of Greenwich', 290-309.
10 Pigafetta, Antonio, 'Diary', quoted in Stanley, Lord, of Alderley (ed.), *The First Voyage round the World* (London, Hakluyt Society, 1874), 161.
11 Dampier, W., *A New Voyage round the World* (London: Argonaut Press, 1927), 255-6.
12 Leigh-Browne, F. S., 'The International Date Line', 305-6.
13 Hellweg, J. F., 'United States Navy time service', *Pub. Astr. Soc. Pacific*, 52, 305 (Feb. 1940).
14 Bigourdan, G., 'Les services horaires de l'observatoire de Paris ...', 30.
15 ibid. 32-3.
16 Bureau des Longitudes, *Conférence internationale de l'heure* (Paris, 1912), D1.
17 Smith, H. M., 'The Bureau International de l'Heure', 29.
18 Willett, William, *The Waste of Daylight* (1907), quoted in full by De Carle, D., *British Time*, 152-7.
19 Wilson, M., *Ninth Astronomer Royal* (Cambridge, 1951), 201.
20 Bigourdan, G., 'Le jour et ses divisions. . .', B72-3.
21 Esclangon F., 'La Distribution téléphonique de l'heure et l'horloge parlante de l'observatoire de Paris', *Annuaire du Bureau des Longitudes pour 1934*, c. 6-11; Parcellier, P., 'L'horloge parlante de l'Observatoire de Paris', *L'Astronomie*, 98, Sept. 1984.
22 Ordnance Survey, *History of the Retriangulation of Great Britain* (1967), 92-101.

Chapter 7 (pp. 163-79)
1 Smith, H. M., 'The determination of time and frequency', *Proc. IEE*, 98, II, 62 (April 1951), 147.
2 Spencer Jones, Sir Harold, 'The Earth as a Timekeeper', *Proc. R. Inst. GB*, xxxiv, 157 (1950), 553.
3 *Trans. int. astr. Un.*, x (1960), 489.
4 'Procès-verbaux des Séances' in *Comité International des Poids et Mesures*, 2ᵉ série, Tome xxv (Paris, 1957), p. 77. (author's trans.)
5 Blair, B. E. (ed.), *Time and Frequency: Theory and Fundamentals* 93-5.
6 *Comptes-rendus des Séances de Ia Treizième Conférence Générale des Poids et Mesures* (Paris, 1968), Resolution 1, p.103. (Translation in Blair, op. cit., p.11.)
7 Smith, H. M., 'International time and frequency coordination', 479-87.
8 Parker, Charles, 'The Passing of Time', *Astronomy Now*, July 1990, 48-9; McIlroy, J. A., 'The history of the Greenwich Time Signal from 1924 to the present day', *Engineering Science and Education Journal*, Dec. 1993, 281-8.
9 *Comptes-rendus des Séances de la Quinzième Conférence Générale des Poids et Mesures* (Paris, 1976), Resolution 5, p.104.
10 Sadler, D. H., 'Mean Solar Time on the Meridian of Greenwich', 290-309: this gives an excellent account of many of the events related in this chapter.
11 Décret no.78-855 du 9 août 1978, loc. cit. 3080.
12 Seventh Directive 94/21/EC of the European Parliament and of the Council, 30 May 1994, Official Journal No.L 164/1,of 30 June 1994.

Appendix I (pp. 181-5)
1 Edwards, Clinton R., 'Mapping by questionnaire: an early Spanish attempt to determine New World geographical positions', Imago Mundi, XXIII (1969), 21-2.
2 RGO 14/67, f.46ᵛ, where this and other lunar-distance observations taken in 1772 are preserved. They are written in a book of printed forms designed by Maskelyne, the page mentioned being reproduced in Howse, *Nevil Maskelyne*, 95.

Appendix II (pp. 185–9)
1 All the instruments mentioned in this Appendix are described and illustrated in Howse, Derek, *Greenwich Observatory, vol. iii: Its Buildings and Instruments* (London, 1975).
2 Perfect, D. S., 'The PZT of the Royal Greenwich Observatory', *O.N. RAS* (1959), 223-33.
3 Danjon, A., 'L'Astrolabe Impersonnel de l'observatoire de Paris', *Bull. Astron.*, XVIII (1954), 251.
4 Guyot, F., *Histoire de la détermination de l'heure*, 117-19.

Index

196